The Destruction of Palestine Is the Destruction of the Earth

The Destruction of Palestine Is the Destruction of the Earth

Andreas Malm

London • New York

First published by Verso 2024
Parts of this book first published on the Verso
Blog, 8 April and 28 May 2024

1 3 5 7 9 10 8 6 4 2

Verso
UK: 6 Meard Street, London W1F 0EG
US: 207 East 32nd St, New York, NY 10016
versobooks.com

Verso is the imprint of New Left Books

ISBN-13: 978-1-83674-007-0
ISBN-13: 978-1-83674-009-4 (US EBK)
ISBN-13: 978-1-83674-008-7 (UK EBK)

British Library Cataloguing in Publication Data
A catalogue record for this book is available from the British Library

Library of Congress Control Number: 2024946012

Typeset in Sabon MT by Hewer Text UK Ltd, Edinburgh
Printed and bound by CPI Group (UK) Ltd, Croydon CR0 4YY

Contents

Preface: No Limits

What follows in these pages is the manuscript for a lecture given half a year into the Gaza genocide, on 4 April 2024, at the American University of Beirut. As the anniversary of Tufan al-Aqsa – commonly translated as 'al-Aqsa flood', although 'tufan' can also mean deluge or storm; it is the root of the English word 'typhoon' – approaches, one thing is abundantly clear: there are no limits to what the state of Israel can get away with. The most recent data at the time of writing, 17 July 2024, speaks of 38,794 people killed; but these are only the bodies that have reached hospitals. Another estimated 10,000 lie buried under the rubble. Of the identified dead, 16,172 are children; and then 34 more have starved to death under the watch of helpless doctors. The occupation has dug seven mass graves inside seized hospital compounds. From these pits, 520 corpses have been recovered after the soldiers moved out. Two million people have had to leave their homes. Almost all of them – 1,737,524 – have suffered from at least one contagious disease, while living in unimaginably overcrowded tent camps or schools; 162 such shelters for displaced people have been bombed. A total of 150,000 housing units have been completely destroyed, as have 115 schools and universities, 610 mosques, 3 churches, 206 archaeological and heritage sites and agricultural lands, the extent of which is not recorded in this particular tally; and to this must be added all the structures damaged beyond repair.[1] But these are, of course, yesterday's figures, since the work of destruction continues without relent or pause, day after day. Again, there are no limits to what the state of Israel can get away with.

Indeed, much of this genocide has unfolded through a blatant and shrill breaking of limits: early on, hospital compounds were considered off-limits to the occupation army (at least in the eyes of some). The repeated assaults on al-Ahli Hospital in Gaza City in late 2023 caused an outcry (albeit faint). The occupation responded by proceeding to al-Shifa Hospital, besieging and destroying it several times over. It then did the same with the Indonesian hospital, al-Quds Hospital, al-Amal Hospital and so on, until the systematic destruction of the hospitals and the massacring of their patients and staff had become a fully normalised feature of death in Gaza. The first massacre of starving Palestinians waiting for trucks to deliver flour elicited indignation (at least from a few corners). The occupation then promptly reacted by conducting more massacres against exactly the same target, until that limit too had been effaced; and the same with bombs targeting families hunkering down in tents, the posting of film clips of soldiers bragging about blowing up civilian homes and images of them playing with the lingerie of Palestinian women – for every atrocity, for every outrage, any kind of reprimand or censure has been countered with a repetition of the act, to the point where the state of Israel has won again: no one can impose any limits on what they do to the Palestinians. This, of course, is not the first time the state has behaved in this manner. Nor is it the first time a colonial entity has done so. 'The colonist is an exhibitionist. His safety concerns lead him to remind the colonised out loud: "Here I am the master,"' Frantz Fanon writes in *The Wretched of the Earth*.[2] Every suggestion of a limit to the destructive power of the settler-colonial state is received as a questioning of its unlimited mastery, and so it must be met with a frenzy of recidivism. The free unfolding of this dynamic can end only with scorched earth in Gaza and beyond.

There was a moment in the spring of 2024 when the master's master, the United States of America, affected the establishment of a limit. The occupation army had not yet entered Rafah. More than a million Palestinians had been driven into this compressed maze of refugee camps dating to 1948 at the southern tip of the enclave and everyone knew that an invasion would bring the catastrophe to new heights: all these people who had been made refugees three or six or ten times over would have to flee again; there would be a mass slaughter of children; the trickle of food and aid would come to a stop. President Joe Biden said, 'I made it clear that if they go into Rafah – they haven't gone into Rafah yet – if they go into Rafah, I'm not supplying the weapons,' without which an operation of this kind, like the genocide as a whole, could not be executed.[3] Rafah was the 'red line'.[4] Rafah must not be pulverised like the rest of Gaza. 'Even President Joe Biden's staunch support can reach its limits,' commented CNN, under the headline 'Biden's Rafah Warning Sends Immediate Shockwaves through US and Global Politics'.[5] But then, of course, the occupation did roll into Rafah, with its tanks and bulldozers and fighter jets, and expelled the population and conducted the standard series of massacres and methodically flattened the camps. As of this writing, more than 70 per cent of the governorate's infrastructure – water wells, roads, sewers, marketplaces – has been destroyed.[6] And, of course, the weapons from the US did keep flowing as naturally as water down an aqueduct. Benjamin Netanyahu has just given a speech to Congress, interrupted by forty-four standing ovations for statements like:

> This is not a clash of civilizations. It's a clash between barbarism and civilization. It's a clash between those who glorify death and those who sanctify life. For the forces of civilization to triumph, America and Israel must stand together [. . .] Our enemies are your enemies, our fight is your fight, and our victory will be your victory.[7]

He and Biden have again been sitting down for a tête-à-tête, thrashing out the details of cooperation and coordination: the master of the earth will never rein in the master of Palestine.

As exasperating and infuriating as this is – or should be – it is in perfect consonance with developments on another front with which the manuscript that follows is also concerned: climate. There are no limits to how much fossil fuel can be extracted. None have yet been placed on the spoliation of this planet. A fresh report demonstrates that the fossil fuel frenzy of the 2020s is still accelerating. Companies are pouring more money into the production of oil and gas today than at any point since the signing of the Paris Agreement, a document that committed the world to limit global warming to 1.5°C. In 2023, the world reached just that limit. The US responded by issuing a record 758 new licences for oil and gas projects in that year alone, almost as many as the three previous years combined, and 2024 could end up with more. The US now pumps more oil and gas than any country has ever done in history: and the curves keep pointing upwards. Under the four years of the Biden administration, the US handed out 1,453 new licences, one fifth more than under the first Trump administration, half of the global total thus far in the 2020s. The frenzy is led by Anglo-America with its settler-colonial petrostates: the UK, Australia, Canada – but above all the US – plus Norway, five affluent countries together accounting for more than two thirds of the licences this decade.[8] Meanwhile, hurricanes without precedent tear through the Caribbean, floods inundate Brazil, heatwaves sear Asia, the tide of climate suffering in the Global South constantly on the rise. And everyone knows that continued extraction will bring the catastrophe to new heights, and yet the licences keep flowing, with the same compulsive limitlessness as the support for the occupation, the same drive to destruction utterly out of control.

How do we think through the relationship between these two processes? It is with this question the following pages are largely concerned, but they merely scratch the surface. There is nothing here in the way of exhaustive inquiry. The text seeks to approach Palestine as a microcosm of larger processes, focusing on a historical moment in 1840 that I believe has particular importance. Still, the story of what happened then is only told with brevity. There are troves of primary and secondary sources – not least in Arabic – that would have to be plumbed for the whole picture to emerge. Work on other projects has prevented me from giving more than a rough (and lightly referenced) account. The text was first published on the Verso Blog – it appears below with only the most minimal changes – and elicited some objections, concerning not the historical narrative but the positions outlined on the Palestinian resistance and the Israel lobby. After the main text, my responses to these objections follow. The first, concerning the resistance, was likewise published on the Verso Blog and has been slightly extended; the second, on the lobby, appears here for the first time.

I wish to extend my gratitude to Nadia Bou Ali and Ray Brassier at the Center for Arts and Humanities and

Critical Humanities for Liberal Arts, at the American University of Beirut, for inviting me to give this lecture and for their amazing generosity during my stay in the city. Many others must also be thanked. Thanks to everyone at the AUB who engaged with the argument; to Hamza Hamouchene and Amr Khairy for translating the text into Arabic; to Matan Kaminer for the discussions; to Shora Esmailian for everything shared; to Ashok Kumar for the kind of comradeship that makes it possible to breathe; to Marco Espvall (whose family in Deir al-Balah is still alive as of this writing) for the Telegram recommendations; to Sebastian Budgen and Michal Schatz at Verso for their unfailing support; to Hèla Yousfi, who invited me to stay at Dauphine University in Paris, during which time most of the lecture manuscript was written, as well as to other comrades in that eternal capital of radicalism in Europe; to Stefanos Geroulanos, who invited me to stay at the Remarque Institute in New York, a visit which happened to overlap with the pro-Palestine student movement in that city, where the response concerning the resistance was written; to participants at the Historical Materialism conference in New York in late May 2024, where the lobby question was discussed. Special thanks to L., to whom this work is dedicated.

Paris, 25 July 2024

The Destruction of Palestine Is the Destruction of the Earth

We have now clocked half a year of this genocide.[1] Half a year has passed since the resistance launched Tufan al-Aqsa and the occupation responded by declaring and executing genocide. It has been half a year, six months, 184 days of bombs picking off one family after another, one high-rise building after another, one residential district after another, relentlessly, methodically: half a year of the grey bones of children poking out under the rubble, of rows of tiny bodies in white shrouds lined up on the ground, of a mutilated girl hanging from a window as if from a meat hook; half a year of parents bidding farewell to their children with eerie composure, as if the souls of the living had departed with the dead, or in uncontrollable spasms of grief, as if they might never again be able to put one foot in front of the other and take a step on this Earth; half a year of a dozen massacres per day, summary executions, sniping, bulldozers driving over corpses and all the rest, and it just doesn't stop, it goes on and it goes on and it doesn't stop and then it continues and proceeds apace and it won't come to an end and it just doesn't stop. One can go insane with despair watching this from a distance. If one feels that way, then one should try to imagine how the survivors in Gaza feel.

The First Advanced Late Capitalist Genocide

The state of Israel is now committing the worst crime known to humanity, and this particular genocide has some unique characteristics that set it apart from others on the recent record. First of all, from its outset, this genocide has been 'a transnational effort', coordinated and organised by the advanced capitalist countries of the West together with Israel.[2] The US, the UK, Germany, France and most other EU members immediately rushed to participate in the bloodshed, sending arms to the occupation as so many dishes to a banquet, flying over Gaza to share intelligence with the headquarters and pilots, rolling out the diplomatic defences around this state and, as if that were not enough, taking the last crumbs of sustenance out of the Palestinians' hands. Now that they are starving and have only the most minimal assistance from UNRWA to keep them alive, the US and the UK are cutting off that last lifeline too.[3] One could be forgiven for thinking they want the Palestinians to die.

This has been the picture during the first half year of this genocide. So far, it has been one monochrome scene of cooperation. No other genocide since the Holocaust has presented such a picture. From Bangladesh to Guatemala, Sudan to Myanmar, genocides might have been perpetrated with varying degrees of complicity from the capitalist core, but here we are dealing with something qualitatively different. A useful comparison would be with the genocide of the Bosnian Muslims, an event that shaped my own political youth. With an arms embargo, the West denied that people the right to defend themselves; through their retreat from Srebrenica, the Dutch forces knowingly handed over the town to Ratko Mladić; in the four years of the war, the so-called international community stood by as Bosnian Muslims were decimated. But these were, primarily, acts of omission. The West did not arm Republika Srpska with the best bombs from its arsenals. Bill Clinton did not fly in to hug Slobodan Milošević. The slaughter was not accompanied by the constant refrain 'the Serbian nationalists have the right to defend themselves'. What we are seeing now might be the first advanced late capitalist genocide.

I must confess to some naivety here: I had not expected quite this voracious an appetite for Palestinian blood. Of course, I have not been surprised by the behaviour of the occupation. The second thing we said to each other on the morning of 7 October was, 'They will destroy Gaza. They will kill everyone.' The first thing we said in these early hours consisted not so much of words but of cries of jubilation. Those of us who have lived our lives with and through the question of Palestine could not react in any other way to the scenes of the resistance storming the Erez checkpoint: this labyrinth of concrete towers and pens and surveillance systems, this consummate installation of guns and scanners and cameras – certainly the most monstruous monument to the domination of another people I have ever been inside – all of a sudden in the hands of Palestinian fighters who had overpowered the occupation soldiers and torn down their flag. How could we not scream with astonishment and joy? It was the same with the scenes of Palestinians breaking through the fence and the wall and streaming into the lands from which they had been expelled; and with the reports of the resistance seizing the police station in Sderot, the ethnically clean colony built on top of the village of Najd, occupied since 1948.

These were the first reactions I shared with those closest to me. But the second: immense trepidation. We all knew how the state of Israel behaves and what to expect of it. What I, personally, did not fully count on was the extent to which the West would throw itself into the mass killings. Clearly, I should have known better. But whatever the naivety, the events of the past half year have raised anew the question of the nature of this alliance. What, exactly, is it that ties the state of Israel and the rest of the West so closely together? What explains the willingness of countries like the US and the UK to collaborate in this genocide? Why does the American empire share Israel's goal of destroying Palestine? One explanation, still as popular as ever on parts of the left, is the power of the Zionist lobby. I will come back to this.

Modes of Pulverisation

One component of the definition of genocide is the 'physical destruction in whole or in part' of the tar-geted group of people; and in Gaza, a central category is precisely that of physical destruction. Already in the two first months, Gaza was subjected to utter and complete destruction.[4] Already before the end of December, the *Wall Street Journal* reported that the destruction of Gaza equalled or surpassed that of Dresden and other German cities during the Second World War.[5] One of the bravest voices outside of Palestine is Francesca Albanese, the UN special rapporteur on the territories occupied in 1967. She begins her recent report with the observation that 'after five months of military operations, Israel has destroyed Gaza', before going on to detail the 'complete destruction of life-sustaining infrastructure'.[6] The emblematic image is that of a house smashed into pieces and survivors frantically digging through the rubble. If they are lucky, a boy or a girl all covered in dust might be pulled from the mass of debris. The estimate now is that some 12,000 dead bodies remain to be extracted from the pulverised houses of Gaza (this figure was later revised to around 10,000).

While it has never before approached the scale we are

now seeing, this is not exactly the first time the Palestinians have experienced this sort of thing. The script can be found in the 1948 Plan Dalet, where Zionist forces were instructed in the art of 'destroying villages (by setting fire to them, by blowing them up and by planting mines in their rubble)'.[7] During the Nakba, it was commonplace for these forces to invade a village during the night and systematically dynamite one house after another with families still inside them. A peculiarity of the Palestinian experience is that this has never come to an end.[8] The original act of destroying the houses over the heads of their inhabitants is repeated again and again: in the village of al-Majdal in 1950, from which the people were driven into Gaza; in Gaza in 2024; and, in between, through any number of eternal recurrences. To pick just one: Beirut in 1982, described by Liyana Badr in *A Balcony over the Fakihani*, with words that could fit any other occasion:

> I saw piles of concrete, stones, torn clothes scattered about, shattered glass, little pieces of cotton wool, fragments of metal, buildings destroyed or leaning crazily [. . .] White dust smothered the district, and through the gray of the smoke loomed the gutted shells of blocks and the debris of houses razed to the earth [. . .] Everything there was mixed up together. Cars were upside down, papers whirling in the sky. Fire. And smoke. The end of the world.[9]

This is the end of the world that never ends: fresh rubble is always poured over the Palestinians. Destruction is the constitutive experience of Palestinian life because the essence of the Zionist project is the destruction of Palestine.

This time, unlike in 1948 or 1950, however, the destruction of Palestine is playing out against the backdrop of a different but related process of destruction: namely, that of the planet's climate system. Climate breakdown is the process of ecosystems being physically destroyed, from the Arctic to Australia. In our book *The Long Heat: Climate Politics When It's Too Late*, forthcoming from Verso in 2025, Wim Carton and I discuss, in some detail, how rapidly this process is now unfolding. To take but one example, the Amazon is caught up in a spiral of dieback that might end with it becoming a treeless savanna. The Amazon rainforest has been standing for 65 million years. Now, in the span of a few short decades, global warming – together with deforestation, the original form of ecological destruction – is pushing the Amazon towards the tipping point beyond which it will cease to exist. Indeed, much recent research suggests that it is perched on that point at present.[10] If the Amazon were to lose its forest cover – a dizzying thought, but entirely within the realm of a possible near future – it would be a different kind of Nakba. The immediate victims would, of course, be the indigenous and Afro-descendent and other people of the Amazon, some 40 million in all, who would, in the most likely scenario, see fires rip through their forest and turn it into smoke and so live through the end of a world.

Sometimes, this process takes on a remarkable morphological similarity to the events in Gaza, even in geographical proximity. On the night of 11 September last year, less than a month before the start of the genocide, Storm Daniel hit Libya. In the eastern city of Derna, on the shores of the Mediterranean, about 1,000 km from Gaza, people were killed in their sleep. Suddenly a force from the sky destroyed their homes on top of them. Afterwards, reports described how random furniture and body parts poked up through pulverised buildings. 'Corpses still litter the street, and drinkable water is in short supply. The storm has killed whole families.'[11] According to one native of the town, it was 'a catastrophe unlike anything we have ever seen. The residents are searching for the bodies of their loved ones by digging with their hands and simple agricultural tools.'[12] There were Palestinian first responders rushing to the scene; according to one of them: 'The devastation is beyond all imagination [. . .] You walk through the city and see nothing but mud, silt and demolished houses. The smell of corpses is everywhere [. . .] Entire families have been erased from the civil registry [. . .] You see death everywhere.'[13]

During its twenty-four-hour visit, Storm Daniel dropped a load of water, around seventy times the average rainfall for September. Derna was located at the mouth of a river running through a wadi towards the sea, normally within narrow banks, if indeed it ran at all. This was desert country. But now suddenly the river rose, burst through two dams and crashed into Derna, the water, sediment, debris forming a bulldozer that ripped and roared through the city in the middle of the night of 11 September – a force of such speed and violence as to drive structures and streets into the Mediterranean and turn the former town centre into a brownish muddy bog. Using today's refined methodologies of weather attribution, researchers could quickly conclude that the floods had been made fifty times more likely by global warming – mathematical code for the cause of the disaster.[14] Only this warming could have brought about that event. During the preceding summer months, the waters off North Africa had been no less than five and a half degrees warmer than the average from the previous two decades. And warm water holds heat energy that can get packed into a storm like fuel into a missile. Some 11,300 people were killed in one single night by Storm Daniel in Libya – the most intense event of mass killing by climate change so far in the decade, possibly the century.

These scenes formed a striking prefiguration of those that would begin to play out in Gaza twenty-six days later; but there were also direct connections between the places. Because rescue teams in Gaza have long been used to dealing with this kind of destruction, they moved swiftly into Derna to help out. At least a dozen Palestinians who had fled from Gaza to Derna were killed in the floods. One Palestinian, Fayez Abu Amra, told *Reuters*, 'Two catastrophes took place, the catastrophe of the displacement and the storm in Libya' – the Arabic word for catastrophe here, of course, being *Nakba*. So, according to Fayez Abu Amra, the first Nakba was the one of 1948, which drove his family and 800,000 other Palestinians out of their homeland; his family ended up in Mukhayyam Deir al-Balah, a refugee camp, and then some members moved on to get away from the Israeli wars of aggression, to the town of Derna; and then came a second Nakba.[15] Fayez Abu Amra lost several relatives in the storm. He himself survived because he had chosen to stay behind in Deir al-Balah, where mourning tents were erected for the victims. And then, just a few weeks later, came the genocide. God knows if Fayez Abu Amra is still alive.

Now, as we recognise the similarities and entanglements of these processes of destruction, some significant differences also strike the eye. The forces that bombed Derna were of another nature than those bombing Gaza. The anonymous sower of death from the sky in the former case was not an air force but the cumulative saturation of the atmosphere with carbon dioxide. No one had the specific intention to destroy Derna, as the state of Israel has had the express intention to destroy Gaza; there were no army spokesmen announcing a focus on 'maximum damage', no Likud MP howling 'Bring down buildings!! Bomb without distinction!!'[16] When fossil fuel companies extract their goods and put them up for combustion, they do not intend to kill anyone in particular. They know, however, that these commodities will, as a matter of certainty, kill people – it might be people in Libya, or in Congo, or in Bangladesh, or in Peru; it is of no consequence to them.

This is not genocide. In our book *Overshoot: How the World Surrendered to Climate Breakdown*, now out from Verso, Wim and I toy with the term 'paupericide' for what is going on here: the relentless expansion of fossil fuel infrastructure beyond all boundaries for a liveable planet. The initial purpose of the act is not to kill anyone per se. The goal of extracting coal or oil or gas is to make money. Once it becomes fully established that this form of money-making actually kills multitudes, however, the absence of intention begins to fill up. As a corollary of the basic insights of climate science, the knowledge is now more or less universally spread: fossil fuels kill people, randomly, blindly, indiscriminately, with a heavy concentration on poor people in the Global South; and they kill in greater numbers the longer business as usual continues. When the atmosphere is oversaturated with CO_2, the lethality of any additional quantum of CO_2 is high and on the rise. Mass casualties are then an ideologically and mentally processed, de facto accepted result of capital accumulation. 'If you're doing something that hurts somebody, and you know it, you're doing it on purpose,' prosecutor Steve Schleicher said in his closing argument against Derek Chauvin, later convicted for the murder of George Floyd; *mutatis mutandis*, the same applies here.[17] Indeed, the violence of fossil fuel production becomes more lethal and more purposeful with every passing year. Compare this with one bombing in Mukhayyam Jabaliya on 25 October, which killed at least 126 civilians, including 69 children. The stated target of this act was a single Hamas commander. Did the occupation intend to also kill the 126 civilians, or was it just callously indifferent to that kind of mass collateral damage? Intentionality and indifference here blur. So, too, on the climate front – still qualitatively different from that of Palestine; but perhaps the difference is diminishing.

Are there any more specific moments of articulation between the destruction of Palestine and the destruction of the earth? By moments of articulation, I mean points where one process impacts and forms the other, in a reciprocating causation, a dialectic of determination. My answer is yes, indeed, such moments of articulation have linked up in a rather tight sequence for almost two centuries now. To demonstrate this phenomenon, I will go back to the moment when it began: 1840. The events of that year have been a perennial obsession of mine. I have touched upon them here and there, but I have yet to write a coherent account. I began doing this research eleven years ago, towards the end of my PhD, when I wrote *Fossil Capital* and realised that the topic required a study of its own, a sequel to be called *Fossil Empire*. In recent months, I have returned again to this moment in time with a view to developing a *longue durée* analysis of fossil empire in Palestine.

Know It Is in Our Power to Pulverise You

The year 1840 was pivotal in history, for both the Middle East and the climate system. It marked the first time the British Empire deployed steamboats in a major war. Steam power was the technology through which dependence on fossil fuels came into being: steam engines ran on coal, and it was their diffusion through the industries of Britain that turned this into the first fossil economy. But steam power would never have made an imprint on the climate had it stayed inside the British Isles. Only by exporting it to the rest of the world and drawing humanity into the spiral of large-scale fossil fuel combustion did Britain change the fate of this planet: the globalisation of steam was the necessary ignition. The key to this ignition, in turn, was the deployment of steamboats in war. It was through the projection of violence that Britain integrated other countries into the strange kind of economy it had created – by turning fossil capital, we might say, into fossil empire.

At this time, Britain was the largest empire the world had ever seen, built on naval supremacy, hitherto founded on the traditional motive power of wind. But in the 1820s, the Royal Navy began to consider steam propulsion – burning coal, that is, instead of sailing with the wind; wind being a 'renewable' source as we would call it today, inexhaustible, cheap, indeed free of cost, but with well-known limitations. The captains could not take for granted that it would blow as they wished. On the battlefield, ships might be held back by calms, driven away from their targets by gusts and gales, or allowed to advance only slowly. Freaks of the wind could give the enemy opportunities to slip away, regroup, hit back. In military action, when the mobilisation of energy was most urgently needed, wind was an unreliable force. Steam obeyed another logic. It drew on a source of energy that had no relation to weather conditions or ocean currents: coal came from the stock underground, a legacy of photosynthesis hundreds of millions of years old; once brought above ground, it could be burnt at whatever point and whatever moment the owner demanded. The striking force of a steamer could be summoned at will. A fleet of such vessels could be arranged just as the captains wanted it – cannons pointed, troops landed, enemies chased down no matter how the wind blew. Such freedom was particularly valued by admiral Charles Napier, the most energetic champion of steam in the Royal Navy, who summed up the case pithily: 'Steamers make the wind always fair'; or 'Steam has gained such a complete conquest over the elements, it appears to me that we are now in possession of all that was required to make maritime war perfect.'[18] The conquest of the elements was, ultimately, a function of the spatio-temporal profile of fossil fuels: their detachment from space and time on the surface of the earth promised to liberate the empire from the coordinates in which boats had navigated since time immemorial.

The first time Napier got to practise this perfection was in 1840, on the shores of Lebanon and Palestine. In that year, Britain went to war against Muhammed Ali. Ali was the pasha of Egypt, nominally serving under the Ottoman Empire but in truth the ruler of his own realm and at war with the sultan. Ali's forces had fanned out from Egypt and conquered the Hijaz and the Levant and formed an Arab proto-empire, on a collision course with the Porte and London. Ali's rise threatened to bring down the Ottoman Empire, whose stability and integrity Britain, at this point in time, regarded as a strategic counterbalance to Russia. If the Ottoman Empire disintegrated, Russia might expand south and east towards the crown colony of India, so Britain wanted to prop it up. Inter-imperialist rivalry, we might say, prompted Britain to intervene against Ali. So did, no less importantly, the dynamics of capitalist development inside Britain itself. The cotton industry was its spearhead, but in the 1830s, it had run so far ahead of every other branch of manufacturing as to suffer a crisis of overproduction: there were mountains of excess cotton thread and fabric coming out of the factories. Sources of demand were insufficient to absorb them all. Britain was therefore desperate for new export markets; and thankfully, in 1838, the Ottoman Empire agreed to a fabulously advantageous free trade agreement, known as the Balta Liman Treaty. This would open the territories under the sultan's control to essentially unlimited British exports. The problem, however, was that ever more of these territories were passing into the control of Muhammed Ali, who pursued the opposite economic policy: import substitution. He built his own cotton factories in Egypt. By the late 1830s, they had grown into the largest industry of its kind outside Europe and the US. Ali would have none of the British free trade: he put in place tariffs and monopolies and other protective barriers around his cotton industry and promoted it so effectively that it could make incursions into markets, hitherto dominated by Britain, as far away as in India itself.

Britain hated this. And no one hated it with more fervour than Lord Palmerston, the foreign secretary and chief architect of the British Empire in the middle of the nineteenth century. He would blurt out: 'The best thing Mehemet could do would be to destroy all his manufactures and throw his machinery into the Nile.'[19] He and the rest of the British government considered Ali's refusal to accept the Balta Liman Treaty a casus belli. Free trade had to be forced onto Ali and all the Arab lands he ruled. Otherwise the British cotton industry would remain suffocated, without the outlets it needed to keep expanding, potentially choked even further by this Egyptian upstart. Lord Palmerston did not conceal his foreign policy principles. 'It was the duty of the Government to open new channels for the commerce of the country'; his 'great object' in 'every quarter of the world' was to prise open lands for trade, and this committed him to an all-out confrontation with Ali.[20] He became obsessed with 'the Eastern question'. 'For my own part, I hate Mehemet Ali, whom I consider as nothing but an arrogant barbarian,' Palmerston wrote in 1839. 'I look upon his boasted civilization of Egypt as the arrantest humbug.'[21] London grew more belligerent by the month. 'Know', the consul-general in Alexandria warned the pasha, 'it is in the power of England to *pulverize* you.'[22] 'We must strike at once rapidly and well' was the advice sent home by Lord Ponsonby, the ambassador in Istanbul, and 'the whole tottering fabric of what is ridiculously called the Arab Nationality will tumble to pieces.'[23] With such words ringing through the corridors of Whitehall, Lord Palmerston ordered the Royal Navy to assemble its best steamboats. In the late summer of 1840, a state-of-the-art squadron under the command of Napier set off for the town of Beirut.

The Pulverisation of Akka

Napier's favourite vessel was the *Gorgon*. Propelled by a 350-horsepower steam engine, with room for 380 tons of coal, 1,600 soldiers and six guns, it was 'the first true fighting steamship', marking 'a new era'.[24] Napier took the *Gorgon* to reconnoitre the area around Beirut, running up and down the coast as he saw fit, in splendid disregard of the weather – but he did send out a pressing request to his fellow officers: 'You must send me coal vessels here *at all costs*, because steamers without coal are useless.'[25] On 9 September, the bombardment of Beirut commenced. *Gorgon* and three other steamers took the lead, a further fifteen sailing ships arrayed around them. Their funnels spewing smoke, the steamers had a distinctive ability to circle the bay of Beirut freely and harass the Egyptian defenders, commanded by Ali's son Ibrahim Pasha. Other targets appear to have been hit as well. After a day of particularly severe shelling, 11 September, the local general sent a letter of accusation to the British fleet:

> For the sake of killing five of my soldiers, you have ruined and brought families into desolation; you have killed women, a tender infant and its mother, an old man, two unfortunate peasants, and doubtless, many others whose names have not yet reached me [. . .] Your fire, I say, became more vigorous and destructive for the unfortunate peasants rather than for my soldiers. You appear decided to make yourselves masters of the town.[26]

Some sources from within Beirut claimed that around 1,000 people were killed in the bombardment, their bodies strewn about the streets. The crew on a US cruiser reported that 'all the buildings, both private and public, were in a heap of ruins, the English fleet were firing upon the few buildings remaining, and were determined not to leave one stone upon another, and the town presents a scene of havoc and destruction'.[27]

After this feat, the steamers chased Ibrahim Pasha's troops along the coast. From Latakia in the north via Trablus and Sur to Haifa in the south, their positions fell like dominoes, the defenders withdrawing under the relentless, unpredictable attacks. 'Steam gives us a great superiority, and we shall keep them moving,' Napier exulted. 'Ibrahim must march very quick if he could beat steam.'[28] A gratified Lord Palmerston followed the news from the frontline, rapidly dispatched with steam packets to London, and wrote back: 'The more force can be accumulated in Syria the better.'[29] Next he ordered an attack on the Palestinian town of Akka. Everyone knew that the decisive battle would take place there. Akka had, famously, held out for half a year against Napoleon in 1799, and then again for half a year in 1831, when Ibrahim Pasha laid siege to it. Since then, the Egyptians had repaired the walls of the old crusader capital, armed its ramparts with heavy guns and garrisoned it with thousands of troops, reinforcing the standing of Akka as by far the sturdiest fortress on the Levantine coast. A major depot, it was filled to the brim with weapons and ammunitions, most of them in a central magazine. It was also a thriving town with a civilian population that had nothing to do with military affairs.

On 1 November 1840, the *Gorgon* and the other three steamers appeared off the coast of Akka. They were alone; the sailing ships had been delayed by light winds. Napier called on the Egyptians to surrender. When they refused, the bombardment began. One report described the action:

> The service of steam-ships in war was thus shewn: the steam division of the Allies having arrived in the Bay, immediately commenced throwing shot and shells into the town, which must have annoyed the garrison very much; as, although they returned a

> very brisk fire, *from the steamers constantly shifting their positions*, it was harmless.[30]

On the evening of 2 November, the wind-powered remainder of the fleet arrived. A proper line of battle was arranged. The special mobility of the new mode of propulsion would be fully utilised, the steamers forming the central prong of the assault:

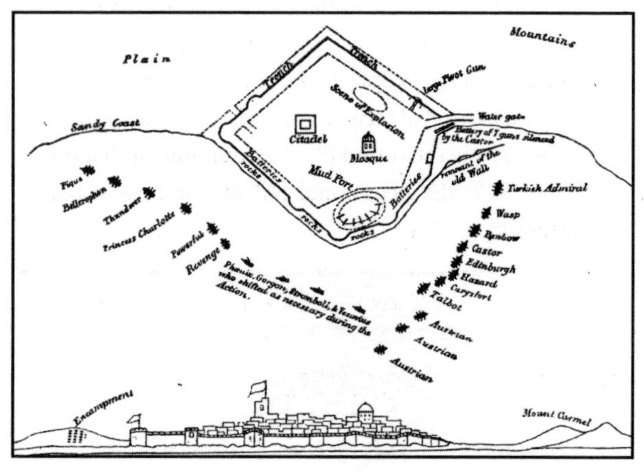

'Plan of the Battle of Acre'. *Source*: W. P. Hunter, *Narrative of the Late Expedition to Syria, vol. I*, 263

In the afternoon of 3 November, the steamers resumed the pounding of Akka and the other ships joined in what was, according to Napier, 'a tremendous fire'.[31] The defenders lobbed back their own shots. After two and a half hours, a deafening detonation ripped through the battlefield. From within Akka, 'a mass of fire and smoke suddenly ascended like a volcano into the sky, immediately followed by a shower of materials of all kinds, that had been carried up by its force. The smoke rested for a few moments like an immense black dome, obscuring everything', read one of many accounts of the event, and further:

> The dreadful crash was heard far above the tumult of the assault, and was immediately succeeded by a most awful pause. The firing on both sides was suddenly suspended, and for a few minutes nothing broke the fearful silence but the echoes of the mountains repeating the sound like the rumbling of distant thunder, and the occasional fall of some tottering building.[32]

John Frederick Warre, 'The Bombardment and Capture of St Jean D'Acre,' (1841)

Akka's great powder magazine had been hit by a shell. The *Gorgon* was dubbed the hero of the strike. In the confident words of one British captain, the 'magazine blew up in consequence of a well directed shell from the "Gorgon" steam-frigate'.[33] We cannot rule out that it was an accidental hit, but the British were clearly aware of the position of the magazine. Relaying fresh intelligence, Lord Minto, the highest commander of the Royal Navy, informed the command on the ground that 'there is much powder stored about very insecurely at Acre' and pointed it out as a suitable target, in a letter signed on 7 October.[34]

Whatever the exact degree of intentionality, the results of the strike from the first true fighting steamship are not in doubt. The Palestinian town of Akka turned into a mass of rubble. 'Two entire regiments', said a report to Lord Palmerston, 'were annihilated, and every living creature within the area of 60,000 square yards ceased to exist; the loss of life being variously computed from 1,200 to 2,000 persons.'[35] As night fell on 3 November, the few surviving Arab soldiers evacuated their last positions in Akka. When the British troops entered the town the next day, they were greeted by utter devastation. Here is one portrait:

> Corpses of men, women, and children, blackened by the explosion of the magazine, and mutilated, in the most horrid manner, by the cannon shot, lay every where about, half buried among the ruins of the homes and fortifications: women were searching for

> the bodies of their husbands, children for their fathers.[36]

In a letter home to his wife, Charles Napier himself expressed unease and perhaps a pang of guilt. 'I went on shore at Acre to see the havoc we have occasioned, and witnessed a sight that never can be effaced from my memory, and makes me at this time even almost shudder to think of it.' He saw hundreds of dead and dying lying in the ruins; 'the beach for half a mile on each side was strewed with bodies'; after some days, the corpses 'infected the air with an effluvium that was truly horrid'.[37] Even in his official account of the *War in Syria*, Napier admitted that 'nothing could be more shocking than to see the miserable wretches, sick and wounded, in all parts of this devoted town, which was almost entirely pulverized'.[38] The British seemed taken aback by the scale of the destruction they had wrought. In a letter to Lord Minto, another admiral wrote, 'I cannot describe to your Lordship the utter destruction of the works & the town from the fire of our ships.'[39] A midshipman from one of the smaller steamers spoke of hands, arms and toes sticking out of the rubble.[40]

Barely remembered today, this event inspired intense fascination in early Victorian Britain. The fortress that held out for half a year against Napoleon went down in less than three days under the blows of the steamships – on the more popular count, less than three *hours* of concentrated bombing on 3 November. It was a sublime, awe-inspiring, miraculous manifestation of the power of Britain in general and steam in particular, rendered in a series of paintings – such as the one by H. Winkles reproduced here, where a steamboat, possibly the *Gorgon*, is pointing straight into Akka, its column of smoke communicating with the tremendous eruption from the magazine behind the walls and the minarets: coal on fire, town on fire.

H. Winkles, 'Bombardment of St Jean D'Acre, by Admiral Sir Charles Napier, 3 November 1840' (1840)

In another lithograph below, presenting the scene from the perspective of the Arab defenders, the smoke from a steamboat likewise rises in the centre, while to the left the whole town is blown sky-high.

The explosion was the centrepiece of the action, but it went further. The steamers made use of their ability to manoeuvre freely in the waters near the walls of Akka, standing as close as 40 metres away when firing their projectiles and then driving back out when needed. The bombardment could be more precise and more devastating thanks to steam power, and it went on for almost three days before the explosion. Did the British use this overwhelming power to target Ibrahim Pasha's forces with the utmost precision? In the most detailed recent

Schranz Brothers, 'Bombardment of St. Jean d'Acre' (1841)

reconstruction of the attack, four Israeli researchers write: 'The bombardment was rather aimed at the town itself [. . .] In fact, the object of the bombardment was to compel the garrison to surrender, not by the injury which it might have sustained, but by the killing and misery which it inflicted upon non-combatants.'[41] We might recognise this kind of strategic thinking. Yet another admiral described how it worked: 'Every shot that cleared the walls smashed the tops of houses, hurling walls and stones down on the heads of people below [. . .] there was no refuge anywhere.'[42]

Whatever compunctions the disembarking men may have felt, back home in Whitehall their joy knew no bounds. Lord Palmerston congratulated the Royal Navy on capturing Akka and securing 'the operation of the commercial treaties'.[43] The road to free trade in the Middle East had been cleared. This was the great achievement of the steamers, widely praised for their efficiency: they 'continually shifted their positions during the action, and threw in shot and shells, whenever they saw the most effectual points for doing execution', observed one report, noticing that 'it is rather remarkable that not one of the four Steam ships had a single man either killed or wounded'.[44] If the men went through the action without a scratch, however, another resource was nearly exhausted: fuel. After the battle, not one of the four steamers had more than one day's supply onboard. Practically all the stored coal had been burnt in the pulverisation of Akka.

The fall of the town determined the outcome of the war in one stroke. Ibrahim Pasha's forces collapsed and beat a disorderly retreat through the coastal plains of Palestine. The steamers continued to harass them, landing at Jaffa and hovering off Gaza. On land, infantry troops moved into Gaza in January 1841, to ensure 'the destruction of the enemy's provisions' – the first time British-led forces occupied this corner of Palestine, if only for a brief moment.[45]

While the British held Gaza and mapped it and destroyed its stores of food – presumably only to deny the Egyptian army its provisions – scattered columns of demoralised, thirsty, hungry soldiers drifted through the desert back into Egypt: less than one fourth of the army Ibrahim had commanded at the outbreak of the war. Before their arrival, Napier steamed onwards to the port of Alexandria, where he threatened to subject that town to the same treatment as Akka unless Muhammed Ali accept all British demands. Ali asked to at least keep the province of Palestine; but again, Napier warned that he would 'lay Alexandria in ashes'.[46] That took Palestine off the table. By the same means, Napier pressed for an immediate implementation of the Balta Liman Treaty in Egypt. Ali caved in on this point, too.

Thus did Britain destroy the Arab proto-empire by means of steam. From Beirut to Alexandria, it was the steamers of the Royal Navy that formed the vanguard of the victory, more expert than their wind-powered partners in every manoeuvre that profited from mobility in space. In an article on the 'Iron War Steamers', the *Manchester Guardian* quoted an anonymous letter from a British subject in Alexandria:

> So much has been done, of late, in the Levant by steam, that everybody is now alive to its capabilities

> as an element either of war or peace, and is ready to ask 'What will it do next?' Ibrahim Pasha can only account for his loss of the coast of Syria in a week by confessing that 'the steam boats conveyed the enemy here, there, and everywhere, so suddenly that it would have required wings to keep up with them! One might as well think of fighting with a genie!'[47]

This power derived from fossil fuels: steam allowed the admirals and captains to plug their boats into a current from the past, a source of energy external to the space and time of the actual battle, through which the ships could therefore shoot as though they had wings. Britain's military superiority was radically enhanced by the ability to mobilise the stock of energy as a force for running the enemy down. Or, as the *Observer* noticed, with reference to Palestine: 'Steam, even now, almost realizes the idea of military omnipotence and military omnipresence; it is everywhere, and there is no withstanding it.'[48] Britain was ready to project the power of fossil fuels across the globe, having proved its mettle in Palestine.

Egypt Subordinated

The country whose fate was most immediately sealed by these events was Egypt. Muhammed Ali's cotton industry crumbled virtually overnight. When free trade was extended to his shrinking realm, the factories on the Nile could not hold out against the British exports, and the reason for this is fairly straightforward: Egypt had no modern prime movers. It didn't have water power, because the Nile is a slow-flowing river with an almost imperceptible gradient, devoid of rapids and falls. Nor did it have steam power. Instead, Egyptian manufacturing ran overwhelmingly on animate power – oxen or mules or even human muscles impelling machines. These were sorely deficient sources of energy compared to steam engines. They were weak, uneven, disorderly. Why, then, did not Muhammed Ali adopt steam? He wanted nothing more. Closely attuned to the trends of capitalist industry, he developed, from the 1820s onwards, a preoccupation with steam and coal bordering on fixation. He knew that he could stand up to Britain only by copying it, in foundries and factories and on the seas, in economic competition and warfare alike. 'The English have made many great discoveries, but the best of their discoveries is that of steam navigation,' he would tell Lord Palmerston's emissary.[49]

But steam demanded fuel. Ali did not possess any coal reserves. He was acutely aware of this problem, so much so that he sent expeditions into Upper Egypt and Sudan and beyond to try to locate seams of coal. My PhD student Amr Khairy recently defended his dissertation *Egypt Ignited: How Steam Power Arrived on the Nile and Integrated Egypt into Industrial Capitalism (1820s–76)*. There he shows how the quest for coal drove the imperial expansion of Muhammed Ali. One of his motives for conquering Syria was the reports of coal in Mount Lebanon. And, indeed, coal could be dug out of the hills from under the Druze and the Maronites: in 1837, the Egyptians managed to extract a volume equal to 2.5 per cent of total British output. Apparently, this Lebanese coal was of inferior quality, expensive, evidently not enough to power a shift to steam in the factories of Cairo before the British struck them down. The nascent coal industry in Mount Lebanon also generated trouble for Ali. People were forced into the mines and abhorred the labour to the extent they rose up against Ibrahim Pasha's forces in 1840; and this uprising was exploited by the British for their own political purposes. The revolt against the coal dreams of Ali contributed to his downfall. His project was to create a fossil empire in the Arab lands; like all builders of empire, he was a ruthless tyrant (in 1834, the people of Nablus revolted against him). In the end, the project came to grief, largely because Ali failed to establish proper coal reserves as a foundation of empire. One can only speculate about what would have happened had the Turkish coal supplies, which we know today to be extensive, fallen into his hands. Shortly after the war in 1840, a declining Muhammed Ali exclaimed to a British visitor, 'Coal! Coal! Coal! That is the one thing needful for me.'[50]

In the 1830s, Egypt balanced on the edge between core and periphery. It embarked on a precocious industrialisation, for a moment the leading 'emerging economy', as it would be called today, outside Europe and the US. But this was a moment in time when access to steam power, and the coal that fuelled it, determined the fate of nations: without this ticket, and with a rough kick from above, Egypt fell down the stairs. The cotton factories on the Nile soon lay in ruins. Egypt became an important market for British exports, and an even more important source of raw cotton: a country locked into the position of a periphery. After 1840, it underwent the most extreme deindustrialisation experienced anywhere in the nineteenth century. Around 1900, somewhere between 93 and 100 per cent of its exports consisted of one single crop – an unusual degree of specialisation. By dint of Egypt's position in the larger Arab world, this underdevelopment also placed the region as a whole in subordination to the advanced capitalist countries of the West: solidified through the events of 1840, a power relation with very durable results.[51] In *Egypt Ignited*, Amr Khairy continues this story in extraordinary granular detail and demonstrates how Egypt became subsumed under the fossil economy that revolved around Britain – its economy was eventually permeated by coal and steam, but it was coal and steam imported from Britain, used for the production and transportation of raw materials. Fortunately, this book will soon be published, so you can read the full account.[52]

Palestine Delivered

The second country whose fate was cast in stone at this time was Palestine. In 1840, the British Empire first proposed its colonisation by Jews. More precisely, on

25 November, Palmerston wrote to Ponsonby, the ambassador in Istanbul: 'This is a great triumph to us all' – the fall of Akka, a few weeks old – 'especially to you, who always maintained that Mehemet's power would crumble under a European attack.' And then he went on:

> Pray try to do what you can about these Jews; you have no idea to what extent the interest felt about them goes; it would be extremely politic [if we could make] the Sultan give them every encouragement and facility for returning and buying lands in Palestine; and if they were allowed to make use of our consuls & ambassador as the channel of complaint, that is to say, place themselves virtually under our protection, they would come back in considerable numbers, and bring with them much wealth.[53]

Fifty-seven years before the first Zionist congress, seventy-seven years before the Balfour declaration, 107 years before the partition plan, the chief architect of the British Empire near the summit of its power here laid down the formula for the colonisation of Palestine. For some reason, this particular document appears to have never been cited in the entire historiography. But it is all there, encapsulated in a missive sent in the euphoria following the pulverisation of Akka.

The year 1840 saw the first mania for what we now know as the Zionist project. It had been in the making for a few years. As is fairly well known, Britain in the late 1830s saw a wave of Christian Zionism, the doctrine that Jews must be gathered and 'restored' to Palestine, where they will convert to Christianity and precipitate the second coming of Christ and usher in the Last Judgement. The main evangelist of this gospel was the Earl of Shaftesbury, who was related through marriage to Lord Palmerston; he tried to make the most of this family bond, but when he spoke to the foreign secretary, he had to put his religious arguments to one side. Instead, he peppered him with reports about 'the productive powers of the Holy Land' being 'for centuries altogether neglected'. If only Britain resolved to insert the Jews into it, Palestine could be turned into a supplier of raw cotton and a market for manufactured goods, and 'our capitalists might be tempted to invest large sums in machinery & cultivation'.[54] After a dinner with Palmerston on 1 August 1840, the godly but shrewd Shaftesbury noted in his diary: 'I am forced to argue politically, financially, commercially; these considerations strike him home.'[55] But eschatology and empire were not incompatible. Shaftesbury succeeded in getting Britain to open a consulate in Jerusalem in 1838; not by coincidence, this was the same year Britain pushed into the region through the signing of the Balta Liman Treaty. God and Mammon mixed rather well. Lady Palmerston, the wife of the foreign secretary, with whom he apparently formed his opinions, read the fall of Akka through her Bible:

> It *cannot* be an accident that all these things should have so turned out! My impression is that it is the restoration of the Jews and fulfilment of the Prophecies [. . .] It is certainly very curious and Acre seems to have fallen down like the walls of Jericho, and Ibrahim's army dispersed like the countless hosts that were enemies of the Jews, as we see in the Old Testament.[56]

It should be pointed out that this was a wholly gentile, Christian, white Anglo-Saxon fantasy, in which actual Jews living in the Middle East or elsewhere played no active role.

Lord Palmerston himself clearly saw the pulverisation of Akka as a sign not of the end times but of a new era of prosperity. No longer would the cotton industry be cramped by a lack of markets. After what he called 'the prostration of Mehemet Ali', Palmerston restated his general philosophy:

> We must unremittingly endeavour to find in other parts of the world new wants for the produce from our industry. The world is large enough, and the wants of the human race ample enough to afford a demand for all we can manufacture; but it is the business of the government to open and to secure the roads for the merchant.[57]

It was in this scheme that the Jews had a role to play. In another letter – and this document has been cited relatively often – Palmerston told Ponsonby to convince the sultan 'to encourage the Jews to return and settle in Palestine because the wealth which they would bring with them would increase the resources of the Sultan's dominions'; moreover, a Jewish settlement would serve 'as a check upon any future evil designs by Mehemet Ali or his successor'.[58] Throughout the 'Eastern crisis', Palmerston again and again dictated this rationale in letters to his ambassador: a 'return' of the Jews to Palestine would implant 'a great number of wealthy capitalists'; if the Sultan would accept them, he would earn the friendship of 'powerful classes in this country' (the UK, that is); 'the capital and the industry of the Jews would much increase his revenue and add greatly to the strength of his empire'.[59] We can here see a kind of brain scan of imperialist Zionism. Because the Jews would be tied to the metropole, giving Palestine to them would help unfetter capitalist development and prevent the rise of new recalcitrant challengers in the region.

As a measure of just how mainstream this scheme had become, *The Times* ran an article on 17 August, while Charles Napier was running up and down the Lebanese coast on the *Gorgon*, explaining that a Jewish settlement of Palestine would function as 'a breastwork against the further encroachments of lawless tyranny and of social degeneracy' – in short, it would be 'advantageously employed for the interests of civilization in the East'.[60] On the ground, the advance detachments of Zionism were formed by officers in the imperial bureaucracy. Some of them came fresh from the battlefield. A colonel by the name of Churchill – Charles Henry, distant relative of the more famous Winston – commanded the British forces that marched into Damascus in early 1841, assembled various dignitaries in a hall and gave a speech:

> Yes, my friends! there was once a Jewish people! famous in arts and renowned in war. These beautiful plains and valleys, which are now tenanted by the wild and wandering Arab, on which desolation has fixed her iron stamp, once revelled in the luxuriance of their fertile and abundant crops, and resounded with the songs of the daughters of Zion. May the hour of Israel's deliverance be near at hand![61]

This Churchill was well aware that there was no, as he put it, 'strong notion among Europe's Jews to return to Palestine'.[62] The desire of Jews to stay where they lived frustrated him. Equally frustrating, his government stuck to the policy of keeping the Ottoman Empire intact, under British guardianship and custody. He wished to see it broken up, and Jewish colonisation of Palestine would be just the right hammer. In a long letter to Moses Montefiore, president of the Board of Deputies of British Jews, sent from Damascus, where he was installed as consul, Churchill exhorted him to convince his fellow Jews to go to Palestine and perhaps Syria too:

> You would end by obtaining the sovereignty of at least Palestine [. . .] I am perfectly certain that these

> countries must be rescued from the grasp of ignorant and fanatical rulers, that the march of civilisation *must* progress, and its various elements of commercial prosperity *must* be developed. It is needless to observe that such will never be the case under the blundering and decrepit despotism of the Turks or the Egyptians. Syria and Palestine, in a word, must be taken under European protection and governed in the sense and according to the spirit of European administration. It must ultimately come to this.

Churchill envisioned a Jewish entity in Palestine under the protection of Britain and its allies, armed for 'defence against the incursions of Bedouin Arabs'.[63]

Another man who hurried into Palestine at this auspicious moment was George Gawler. Having just relocated from South Australia, where he had been governor, he penned a pamphlet titled *Tranquilization of Syria and the East: Practical Suggestions in Furtherance of the Establishment of Jewish Colonies in Palestine, the Most Sober and Sensible Remedy for the Miseries of Asiatic Turkey*. He travelled in Palestine in the early 1840s and somehow managed to perceive it as '*a fertile country, nine tenths of which lie desolate*'. The land was empty, save for a few 'unlettered and restless Bedawy' now and then encountered in 'deserted cities, and thorn-covered plains'. Solution: 'REPLENISH THE DESERTED TOWNS AND FIELDS OF PALESTINE WITH THE ENERGETIC PEOPLE', the Jews, who would turn it into a flourishing marketplace under the watch of a 'naval force frequently on the coast' – the British steamers, that is.[64] A friend of Palmerston, E. L. Mitford, likewise imagined Palestine as 'barren and desolate'. Jewish colonisation would bring 'blessings on England and be felt in the wretched hearts and homes of the poor manufacturers of Manchester, Birmingham and Glasgow'; of particular importance, it would facilitate fossil-fuelled entrenchment in the region and beyond.[65] An independent Jewish state under British protection would 'place the management of our steam communication entirely in our hands and would place us in a commanding position in the Levant from whence to check the process of encroachment, to overawe open enemies and, if necessary, to repel their advance'.[66] Such was the formula derived from the events of 1840.

This, then, was the moment of conception for two interrelated principles: one, no people exists in Palestine; two, the land must be taken with the force of technology running on fossil fuels. As for the former, contemporary Zionists debate who first came up with the slogan 'a land without a people for a people without a land', but there is a consensus that it happened around the year 1840. Some point to an article Shaftesbury wrote in *The Times* in 1839, where he used the phrase 'Earth without people – people without land', *Earth* without people sounding perhaps slightly more chilling today.[67] Others give the honour to his fellow Christian Zionist Alexander Keith, who went on an expedition to Palestine in 1839 and somehow managed to return with the impression that this was a 'country without a people' crying out for the arrival of 'a people without a country'.[68] The cities and towns of Palestine were 'desolate without an inhabitant'; from Gaza to al-Khalil, all Keith could observe were 'deserted sites and ruined villages, not one of them being inhabited'.[69] But now a miracle had transpired. 'As if commissioned by the Lord,' Keith wrote of Akka, 'a bomb penetrated a magazine of powder stored up for defence, and raised the arsenal in the air, as if to show that the time was come that the last *fortress* in Palestine should *cease*, and strewed it stone by stone upon the ground' – strewed it stone by stone upon the ground – 'as if the times too were not distant when the hands of strangers should find other work, and build up the ruined walls in another form [. . .] Acre fell to the lot of a tribe of Israel.'[70]

It now became a persistent theme of British commentary on Palestine that no people lived there. Shaftesbury informed Palmerston that Jewish colonisation would be 'the cheapest and safest mode of supplying the wastes of those depopulated regions'.[71] The *Morning Post* published a typical article claiming that 'Syria and Palestine are depopulated', voids in which the 'sons' of 'the Arabian wilderness' had failed to 'establish themselves and maintain their nationality'.[72] The year 1840 was here calculated to match a Biblical prophecy of Jewish restoration. Such fusion of eschatology and empire became very much in vogue after Akka, as in perhaps the most peculiar tract to emerge from this moment, an anonymously authored 350-page mishmash of exegesis, realpolitik and steam fetishism called *The Kings of the East*. Here, too, Palestine was said to have 'few' inhabitants, and the fall of Akka was hailed as a divine intervention by means of steam, the pillar of British power.[73] As proof of the metaphysical significance of Akka, the author quoted a first-hand report of how 'the town is a complete mass of ruins: not a house in the place, however small, has escaped the fury of our shot [. . .] Everything bears the most ample witness of the matchless precision of our guns' – praise the Lord: 'thousands of her garrison numbered with the dying and the dead'.[74] Ergo, restoration was imminent. This author alleged that 'the Jews are commencing to return to Judea'.[75]

Two verses of the Bible shed special light on this process. At the opening of the eighteenth chapter of Isaiah, in the King James version, we read: 'Woe to the land shadowing with wings, which is beyond the rivers of Ethiopia: That sendeth ambassadors by the sea, even in vessels of bulrushes upon the waters, saying, Go, ye swift messengers, to a nation scattered and peeled' – what kind of vessels did the prophet talk about here?[76] Patently, he must have had British steamboats in mind. They were the ones sending ambassadors by the sea to open Palestine for the Jews. From this, the author inferred a novel prophecy: Britain 'WILL ISSUE A PROCLAMATION, guaranteeing to all Jews who will return to Syria their protection'.[77]

The mania also crossed the Atlantic to the United States of America. In the weeks before the fall of Akka, one influential, relatively progressive periodical, *The Western Messenger*, knew which way the wind was blowing: 'Now that steam ships ride in the bay of Alexandria, and steamboats break the waters of the Nile, and the roar of steam cars, dashing over railways, is heard, is it not morally certain that the Moslem power has ceased?' The time had come to 'give the Jews possession of Palestine'.[78] They would take and defend the land with military might and send all sorts of profit flowing back to the West. But the first American Zionist of import, who, unlike nearly all of his British counterparts, also happened to be Jewish, was Mordecai Manuel Noah.[79] In 1844, he delivered a *Discourse on the Restoration of the Jews*. He never visited Palestine, but he seems to have learned from British travellers that 'the land is now desolate' – although he did relay that 'olives and olive oil are everywhere found', and wheat and corn and cotton and tobacco grow in the plains and hills, and 'grapes of the largest kind flourish everywhere'. Now 'great and important revolutions' were underway in that land. Noah latched on to the view that the demise of Ali heralded 'the organization of a powerful government in Judea', and he implored the US to take it under its wing.[80]

Noah likewise read Isaiah 18. He deepened the exegesis one notch by reading it directly in Hebrew and finding that the prophet called the vessels *gomey*. This Hebrew word could also mean 'an impetus, a forcible propelling power' – further evidence that the prophet referred to steam. But in Noah's reading, the steam power would be American, not British. 'The land lying beyond the rivers of Ethiopia is America' and the vessels 'our steam vessels', with a divine mission to settle the Jews in Palestine.[81] 'The discovery and application of steam will be found to be a great auxiliary in the promotion of this interesting experiment.' It placed American Jews 'within a few days' travel of Jerusalem. Our Mediterranean and Levant trade, hitherto much neglected, will be revived, affording facilities to reach Palestine from this country direct.'[82] Reverting to the default view that no economic activities were occurring as of now, Noah anticipated that 'the ports of the Mediterranean will be again opened to the busy hum of commerce; the fields will again bear the fruitful harvest'.[83] He looked forward to a future when

> the whole territory surrounding Jerusalem, including the villages Hebron, Safat, Tyre, also Beyroot, Jaffa, and other ports of the Mediterranean, will be occupied by enterprising Jews. The valleys of the Jordan will be filled by agriculturists from the north of Germany, Poland, and Russia. Merchants will occupy the seaports, and the commanding positions within the walls of Jerusalem will be purchased by the wealthy and pious of our brethren.[84]

Some prophecies do come true.

A Structure, Not an Event

What do we make of all this? Here is the first moment of articulation: the moment that ignited the globalisation of steam, through its deployment in war, was also the moment that conceived the Zionist project. But there was no perfect synchrony. Zionism was, as yet, only an idea. No Jewish settlement in Palestine developed in the wake of 1840; strictly speaking, the Palmerstons, Shaftesbury, Churchill, Gawler, Noah and the others all failed. They were ahead of their time by half a century. But when the Zionist movement was eventually assembled, it was a wagon that could be placed on ready-made tracks, laid out by the British Empire after 1840: the dominant classes of the metropole had already constructed the logic of its satellite colony in Palestine, if only as a mental image. Zionism did not take material form in 1840, as the exercise of steam-powered violence did. We might conclude from this that the latter had causal primacy in history. Zionism first existed at the level of superstructure, on the base of the fossil empire.

I do not say this with a pretension to revolutionary discovery. The broad contours of this story can be found in the existing historiography, including the most recent sustained engagement with the period, *Promised Lands: The British and the Ottoman Middle East* by Jonathan Parry. He chronicles how the British thrust into the region by means of steam. 'From the 1830s, steam power', he writes, 'was a valuable aid in intimidating Arabs into appreciating British might.'[85] Beyond the Levant, two Arab lands in particular were subjected to this might: Yemen and Iraq. In 1839, Aden was occupied and annexed as a coaling station for steamboats; in the late 1830s, various experiments with steam communication were launched on the Euphrates. By 1841, when the British had done away with the main obstacle, 'their regional naval supremacy was undisputed. Whether steam might civilise the Arabs was a question for the long term,' Parry coyly adds.[86] He works in the tradition of gentlemanly, tea-sipping British history-writing and so refuses to draw out implications or follow lines; he also studiously ignores political economy and represses the mountains of evidence for how the dynamics of capital accumulation propelled the expansion into the Middle East – evidence of which I have only here provided the tiniest sample. But the attentive reader can make out the narrative arc.

'A large proportion of the things the British ever thought about the Middle East had already been thought by 1854,' states Parry.[87] We can sharpen this and state that a large proportion of the things the British and the Americans ever thought about Palestine had already been thought by 1844. And it began with extreme technological superiority, the penetration of the region by means of cutting-edge fossil-fuelled machines. That kind of subjugation would remain in place up to the present day; what happened in 1840 was no ephemeral intrusion, like Napoleon's campaigns. The British would not let go of the Middle East – they moved only deeper and deeper into it, until in the last decade of the nineteenth century, having occupied Egypt, they had risen so high as to lay the Ottoman Empire low enough for the colonisation of Palestine to get going. All the UK ever did was to share this power with and pass it on to the US. But as the ongoing bombardment of Yemen testifies, the British are still very much there.

A few more words on the dialectic of mind and matter may be in order. There is an odd spiral of reality and fantasy at work in the moment of 1840: the British really did turn one Palestinian town into ruins. Then they started imagining that all of Palestine was one landscape of ruins – desolate, deserted, depopulated; fanciful constructions at best, but rather adequate representations of what Akka seems to have looked like after 3 November. In the next coils of the spiral, the ideational emptying of the land became a precursor to the real thing. 'Earth without people' read the prescription for a Nakba. Ever the pioneers, the British undertook a prefigurative elimination of the Palestinian people. At this moment in time, curiously, Jews still had a position rather symmetrical to the Palestinians: they existed as characters in the plot, but purely in the realm of the imagination. Actual Jews did not count. Jews did not clamour to abandon their homes for Palestine – rather to the contrary, as even one Zionist scholar has noticed, 'British Jewry was opposed to "anything that might seem to impugn its status as 'wholly' English." English Jews could only be embarrassed by the suggestion that they were waiting to go back to Palestine.'[88] Before Zionism was Jewish, it was imperial.

But real Jews would, of course, in time be recruited into the Zionist project, and real Palestinians would be erased from their land. Set in the context of this *longue durée*, the genocide in Gaza does not appear all that accidental. In her report to the UN, Albanese is audacious enough to draw on the school of settler-colonial studies to explain it. She writes: 'Israel's actions have been driven by a genocidal logic integral to its settler-colonial project in Palestine, signalling a tragedy foretold.' Genocidal extermination is the climax of settler colonialism, and in Palestine, from the moment of 1948, 'displacing and erasing the Indigenous Arab presence has been an inevitable part of the forming of Israel as a "Jewish state"'.[89] She is right, of course. But settler colonialism in Palestine never stood on its own feet and never could have. And the tragedy was foretold earlier than by the likes of Yosef Weitz. The Palestinians had already been figuratively spirited out of Palestine 183 years before this genocide; with some interruptions and fits and starts, the materialisation and escalation of the act have been in motion ever since. Consider the words from Isaac Herzog, president of the occupation, adduced by Albanese as one instance of genocidal intent: he affirmed in October and November that his entity fights on behalf of 'all civilized states [. . .] and peoples', against 'a barbarism that has no place in the modern world' – it will 'uproot evil and it will be good for the entire region and the world'.[90] These words could have been put in his mouth by the Anglo-Zionists of 1840.

We might paraphrase the motto of the school of settler-colonial studies and say that imperial support for the Zionist entity is a structure, not an event. The structure was forged by the exceptional power accorded to those armed with fossil fuels and has continued to function that way, as I will now briefly argue, but before I do so, let me point out one last thing about 1840: the account I have given here is sketchy and partial. Most problematically, it relies exclusively on English sources. I do not read Arabic, so I cannot say whether there is an Arabic historiography of 1840. Nor does Parry read Arabic, but he tells us: 'There are many non-English archives that seem not yet to have been fully used by anyone.'[91] Whatever Arabic sources from and about 1840 exist, and whatever they say about this original encounter with the power of steam and the notions of Zionism, they have not yet left a trace in the English literature. Deep research on this moment would begin with some digging outside the metropole.

Steps of Dual Destruction

Now, I will be extremely sweeping and synoptic in what follows. When the first Zionist colonies were built, one could find excited reports in the Western press: 'The Jews who are now going to Palestine take with them the progressive spirit of the century, and before long travellers in that country may hear the steam whistle, and the clatter of machinery, and see all around them the bustle of business instead of the traditional apathy and listlessness of the Orient,' the *National Repository* rejoiced in 1877.[92] When the British Empire occupied Palestine and set about implementing the Balfour Declaration, the fossil fuel of the day was not coal. It was oil.

Promising deposits had been located in the countries bordering the Persian Gulf, and the central industrial project of the Mandate came to be the pipeline that brought crude oil all the way from Iraq, across the northern West Bank and the Galilee, to the refinery in Haifa. The Mandate as such cannot be understood outside the deepening control over the region in the pursuit of oil; and the Mandate used oil to reallocate land from Palestinians to Jews. In his forthcoming *Heat: A History*, a wonderfully rich history of high temperatures and fossil fuels in the Middle East, On Barak shows, among many other things, how the Yishuv wrested citrus production from Palestinians by linking up with the most modern circuits of technology: irrigating their orchards with fossil-fuelled pumps, loading their fruits on lorries, sending them over roads to ports, offloading them onto steamers to the European market – a symbiosis with the fossil empire by which the natives could be squeezed out of their iconic citriculture. The Mandate authorities systematically privileged the building of roads between colonies. Oil-based infrastructure tilted Palestine in the direction of the settlements on the coastal plains and further towards their patrons on the other side of the ocean.

When Zionist forces began to terrorise the Palestinians of Haifa to drive them out of the city, Ilan Pappe writes, 'rivers of ignited oil and fuel [were] sent down the mountainside'.[93] When the top echelons of the US empire discussed whether to throw in their lot with the Zionists during the Nakba, they had oil interests foremost in mind. Some argued that these would be better served by siding with the Arabs. But as Irene L. Gendzier has demonstrated in *Dying to Forget: Oil, Power, Palestine and the Foundations of US Policy in the Middle East*, the government was swayed by the argument that a Palestinian victory would 'increase Arab self-reliance, demands and bargaining power', whereas the establishment of the state of Israel 'would have a soothing effect on the Arabs and make them regain their right sense of proportion'; moreover, 'the Yishuv is a Western progressive factor, which will be a great stimulant to any social progress in the Middle East, which will open new commercial markets'.[94] The American oil companies seem to have converged on the view that control over deposits would be indirectly reinforced by having Israel as an ally in the region. And that is indeed what transpired during the 1950s and '60s, the golden age of the Seven Sisters and Persian Gulf oil. When the US took over the role as the number-one backer of Israel after 1967, the defence of this status quo was the paramount concern. In *The Global Offensive: The United States, the Palestine Liberation Organization, and the Making of the Post–Cold War Order*, Paul Thomas Chamberlin describes how the US regarded Palestinian liberation as a threat to the domination of the Middle East as a whole, with all its invaluable oil reserves. Conversely, 'Israel was fast proving its value as a key strategic asset in the Middle East and a model regional policeman in the Third World.'[95] Proof of this logic came from the event known as Black September, one of the eternal recurrences, depicted in a letter from Yasser Arafat on 22 September 1970: 'Amman is burning for the sixth day [. . .] The bodies of thousands of our people are rotting beneath the rubble.'[96]

All of this, it should now be clear, followed the script first laid down in 1840. If Plan Dalet was a settler-colonial script for the destruction of Palestine from 1948 onwards, it was preceded by – and had its conditions of existence in – the imperialist vision of an entity imposed on the land of Palestine for the protection of the interests of the core: access to raw materials and markets; prevention of subversive projects; buffer zones and counterweights against more distant rivals. In 1840, it was cotton, Muhammed Ali and Tsarist Russia. When the occupation was completed 127 years later, it was petroleum, Third World liberation and the Soviet Union. We are dealing here with an exceedingly deep structure, not an event or two; a ratcheting-up and escalation across two centuries, a worsening and intensification of patterns first developed in the early nineteenth century – also, not coincidentally, the temporal form of global warming itself. I have pointed very briefly and superficially to three further pivotal moments of articulation. In 1917 and after, the British occupation of Palestine was part of the transformation of the Middle East into a foundation for fossil capital, by dint of its oil resources. In 1947 and after, Western support for the new Zionist state was informed by the consummation of that order; in 1967 and after, by its defence. The steps along the way to the destruction of

Palestine were simultaneously steps along the way to that of the earth.

Fossil Israel

If we now jump to the present, we should first consider the role of the state of Israel in the ongoing fossil fuel frenzy. In *Overshoot*, Wim and I show in some detail how the 2020s have so far seen an accelerated expansion of fossil fuel production, just when it had to be reined in and inverted into its opposite – a sustained dismantling – for the world to avoid a warming of more than 1.5°C or 2°C. This expansion doesn't stop: as the *Guardian* reported just the other day, corporations and states are forging ahead with new oil and gas projects in ever-growing volumes.[97] The country leading the expansion is, of course, the US; second on the list is Guyana, but that is only because ExxonMobil has found its most recent treasure trove in its waters. And for the first time, the Israeli state is directly involved. One of the many frontiers of oil and gas extraction is the Levant basin along the coast running from Beirut via Akka to Gaza. Two of the major gas fields discovered here, called Karish and Leviathan, are located in waters claimed by Lebanon. What does the West think of this dispute? In 2015, Germany sold four warships to Israel so it could better defend its gas platforms against any eventualities.[98] Seven years later, in 2022, as the war in Ukraine caused a crisis on the gas market, the state of Israel was for the first time elevated into a fossil fuel exporter of note, supplying Germany and other EU states with gas as well as crude oil from Leviathan and Karish, which came online in October of that year.[99] The year 2022 sealed the high status of Israel in this department.

A year later, Tufan al-Aqsa threw a spanner in the expansion. It posed a direct threat to the Tamar gas platform, which can be seen from northern Gaza on a clear day; in the range of rocket fire, the platform was shut down.[100] A major player on the Tamar field is Chevron. On 9 October, the *New York Times* reported:

> The fierce fighting could slow the pace of energy investment in the region, just as the eastern Mediterranean's prospects as an energy center have gained momentum. Israel used to be one of the few countries in the Middle East without significant discovered petroleum resources. Now, natural gas has become a mainstay of its economy.[101]

The Palestinian resistance could upend this equation. Five weeks after 7 October, however, when most of northern Gaza had been comfortably turned into rubble, Chevron resumed operations at the Tamar gas field.[102] In February, it announced another round of investment to further bolster output.[103] In late October, the day after the ground invasion of Gaza began, the state of Israel awarded twelve licences for the exploration of *new* gas fields – one of the companies picking them up being BP, the very same company that first discovered oil in the Middle East and built the Kirkuk–

Haifa pipeline.[104]

But the feedback loop now goes both ways. Israeli capital has in recent years become a major player in the expansion of oil and gas production in the North Sea. One of the companies based in Tel Aviv and spearheading extraction off Akka, as well as the Shetland Islands, is Ithaca Energy: it owns one of the most destructive carbon bombs planted in the British sector of the North Sea, the Cambo field, and one fifth of another, the Rosebank field, and it hungrily explores for more.[105] When Ithaca entered the London stock exchange in 2022, it was the largest floatation of that year.[106] BP is looking for gas in the waters of Palestine; Ithaca is looking for it in the waters of Britain: the harmony has never been greater. The genocide is unfolding at a time when the state of Israel is more deeply integrated in the primitive accumulation of fossil capital than ever. The Palestinians, on the other hand, have zero stake in that process: no platforms, no rigs, no pipelines, no companies listed on the London Stock Exchange. But Arabs in the UAE and Egypt and Saudi Arabia do, of course. This is the political economy of the Abraham Accords and its expected sequels: a unification of Israeli and Gulf capital in the process of making money by producing oil and gas. This is the political ecology of normalisation: a sacralisation of business as usual that destroys first Palestine and then the earth.

The Combustion of Gaza

The destruction of Gaza is executed by tanks and fighter jets pouring out their projectiles over the land: the Merkavas and the F-16s sending their hellfire over the Palestinians, the rockets and bombs that turn everything into rubble – but only after the explosive force of fossil fuel combustion has put them on the right trajectory. All these military vehicles run on petroleum. So do the supply flights from the US, the Boeings that ferry the missiles over the permanent airbridge. An early, provisional, conservative analysis found that emissions caused during the first sixty days of the war equalled annual emissions of between twenty and thirty-three low-emitting countries: a sudden spike, a plume of CO_2 rising over the debris of Gaza.[107] If I repeat the point here, it is because the cycle is self-generating, only growing in scale and size: Western forces pulverise the living quarters of Palestine by mobilising the boundless capacity for destruction only fossil fuels can give.

It is easy to forget just how central military violence has been and remains to business as usual. More than 5 per cent of annual CO_2 emissions stem from the militaries around the world. We often talk about flying and how bad it is for the climate, and it is bad, but civil aviation accounts for about 3 per cent of the total. And the 5 per cent that comes from militaries precede actual war: these are peacetime emissions, made in the process of maintaining the logistical apparatuses and fighting capacities of armies before they go to war. When they do go into battle, the fuel is set on fire and the bombs rain down in bursts of concentrated additional emissions. The US, of course, is at the centre of all this. The emissions from the occupation army during the war on Gaza might be counted as just another category of American emissions. The US outweighs every other country; indeed, as Neta C. Crawford notes, 'the US military is the single largest institutional fossil fuel user in the world and thus the world's single largest greenhouse gas emitter.'[108] In her book *The Pentagon, Climate Change, and War*, she brilliantly charts the development of what she calls 'the deep cycle'. The militaries of first the UK and then the US found coal, followed by oil, to be indispensable for waging war: for manufacturing weapons, transporting soldiers into the battlefield, providing mobility once engaged, bringing firepower to bear on the enemy. By basing its operations on fossil fuels, the US military contributed to their spread throughout the economy; and when both the military and economy were thoroughly dependent on them, the protection of this essential commodity itself became an imperative of war. No part of the world has been so deeply formed and scarred by this cycle as the Middle East. Although Palestine is at its centre, the devastation clearly extends to other countries too: think only of Iraq and Yemen.

Against the Lobby Theory

Let us then revisit the question of the nature of the alliance and briefly reconsider the theory of the lobby. It says, in short, the following: the Zionist lobby in the US has amassed so much financial, electoral and media power as to hold American politics in an iron grip. Through its machinations and manipulations, it has compelled the US to support Israel, despite this not being in the real, rational, material interests of the country. The US backs Israel for reasons of domestic politics, which distort the preferences and standing of the US in the international arena. The theory is, of course, based on the work of John Mearsheimer, a man of the US military, a so-called realist with no ideological affinity to the left. I find the avid reception of his work on parts of the left rather surprising. Space does not allow for a comprehensive critique of either Mearsheimer or his echoes on the left: here I shall only point out some of the problems in one representative rendition of the theory.

Married to Another Man: Israel's Dilemma in Palestine by Ghada Karmi is a widely read and average statement of the case for Palestine in the early twenty-first century. She correctly observes that, for Palestinians, understanding the nature of the alliance between the US and Israel is 'no intellectual game but a matter of life and death'.[109] She poses two alternative explanations: 'Was US policy so controlled by Israel and its supporters that it was they who primarily dictated it, or was Israel but the imperialist arm of America (and the West) in the

Middle East?' and she comes down firmly on the side of the former.[110] She goes on an unclear rant about Jews in the media and Hollywood and concludes that this country is the victim of 'a foreign state's penetration into the US system'. A counterfactual typical for the theory is constructed: 'Had the situation been one of rational, pragmatic common sense, where the facts could be examined and the logical conclusions drawn, then the American national interest would ultimately have prevailed over the forces working on Israel's behalf.'[111]

If only the US were free to pick the policy that would best serve its interests, it would dump Israel. But the Zionist lobby denies the state that freedom. This distortionist explanation applies not only to Palestine, but to the region in its entirety. Everything the US does in the Middle East is dictated by Israel against its true interests. 'The real motivation for the invasion of Iraq', we learn, was 'the desire to protect the Jewish state' foisted onto the US; there were no weapons of mass destruction, no al-Qaeda, no terrorism in Iraq, so 'it must have been Israel's security that was the motive for attacking Iraq, in the absence of any other'. This is a double non sequitur. It does not follow from the absence of these official casus belli that the real reason must have been the security of Israel; but it does follow from their absence that the security of Israel was not threatened by Saddam Hussein. Karmi wants us to believe that Israel was out for the Iraqi oil and sent businessmen and advisors and intelligence agents into the country, whereas the US itself, possessing none of these aggressive drives, was dragged passively into the war by the lobby. We are asked to believe, in other words, that the most powerful empire in the history of the world has no interests and performs no aggressions of its own in the Middle East. It is the same with Syria and Iran, Karmi tells us: what the US does to those countries, it does slavishly on behalf of Israel.

Despite Karmi mentioning Shaftesbury and Palmerston in passing, without any serious historical analysis, she aspires to this being a chronologically accurate explanation: 'It was the arrival of Israel and the powerful lobbies working on its behalf that forced successive US administrations to find a way for it in their foreign policies.' So, first came Israel and the Zionist lobby and then the empire was coerced into obeying them.[112] I think we can safely conclude that even the limited evidence presented here should be enough to rebut this theory. The historical evidence points to the validity of the contrary explanation. I think Sayyed Hassan Nasrallah gets it right when he says:

> There is a misconception prevalent in the Arab world regarding Israeli–US relations. We keep hearing this lie about the Zionist lobby – that the Jews rule America and are the real decision-makers, and so on. No. America itself is the decision-maker. In America, you have the major corporations. You have a trinity of the oil companies, the weapons industry, and the so-called 'Christian Zionism'. The decision-making is in the hands of this alliance. Israel used to be a

> tool at the hands of the British, and now it is a tool in the hands of America.[113]

This, of course, is the classic position taken by the Arab left and the keenest analyses from the Palestinian resistance. In *Strategy for the Liberation of Palestine*, the foundational document of the Popular Front for the Liberation of Palestine (PFLP) from 1969, the enemy is defined as a dialectical unity of global imperialism and local settler colonialism: victories of the latter are 'fundamental matters for the interests' of the former. The entity is an imperialist 'base on our land and is being used to stem the tide of revolution, to ensure our continued subjection and to maintain the process of pillage and exploitation'; Zionism is 'an aggressive racial movement connected with imperialism, which has exploited the sufferings of the Jews as a stepping stone for the promotion of its interests [. . .] in this part of the world that possesses rich resources and provides a bridgehead into the countries of Africa and Asia'.[114] This is the antithesis of the lobby theory. It can also be found in the best writings of Islamic Jihad, such as its political document from 2018, where we read that 'the Zionist project is the project of a settler-colonial invasion', but it is 'based on the organic link with the forces of Western colonialism, which worked to get rid of the Jews and to solve the "Jewish problem" in Europe by planting an entity for the Jews in Palestine'. The persistence of that entity 'is essentially related to the role assigned to it. It is a tool' – pace Karmi, a tool – 'for the project of colonial domination' and 'derives all of its material and moral strength from the strength and capabilities of the West, in particular the United States of America'.[115] Fathi al-Shiqaqi took the outlines of this analysis from none other than Izz al-Din al-Qassam. In the early 1930s, he opposed the Palestinian leaders who 'regarded it as necessary to reason with Britain to make it stand with us against the Jews, thus forgetting and ignoring that Zionism is no more than another imperialist face of Britain'.[116]

Unlike the distortionist theory of the lobby, the instrumentalist theory of empire and entity is confirmed by evidence from the deep past, as well as from the recent past and the present: Joe Biden could have stepped out of the pages of a Jabha or Jihad document. In 1986, this future president told Congress:

> There is no apology to be made for Israel. None! Israel is the best 3-billion-dollar investment we make. Were there not an Israel, the United States of America would have to *invent* an Israel to protect our interests in the region. The United States would have to go out and *invent* an Israel.[117]

It couldn't be much clearer than that, nor more in line with the historical record of invention. In 2007, Biden reaffirmed that 'Israel is the single greatest strength America has in the Middle East [. . .] Imagine our circumstance in the world, were there no Israel'; and then in 2010, he repeated that 'there's simply no space between the United States and Israel'; but his most oft-repeated line was the one about having to invent Israel if it didn't exist – most recently, he said this again in July 2023, in a meeting with Isaac Herzog in the White House.[118] That was three months before the genocide began.

The left should make a sharp break with the lobby theory. That is not to suggest that we have a complete understanding of the relation between empire and entity – to the contrary, I think the remarkable thing here is that we do *not* have, for instance, and correct me if I'm wrong, a single good book in English about how the structure works at present. Where is the US empire going? What is it doing in the Middle East? How does the state of Israel fit in? – I don't think we possess anything like a comprehensive, up-to-date, empirically and theoretically grounded set of answers to these questions, because the hard work of researching and thinking is yet to be done. There is a debilitating deficit of *au courant* analysis of American and other Western imperialisms, perhaps because the left has found it a slightly embarrassing pursuit, too reminiscent of orthodox Leninism and campism and other sources of shame. I am personally unqualified to fill this gap, but let me just throw out the hypothesis that the share value of Israel as an investment rises proportionately to the challenge from Russia and China. When inter-imperialist rivalry intensifies again, in the 2020s much as in the 1830s or the 1910s, the entity becomes an invaluable asset. From the first moments of Tufan al-Aqsa, it was clear that a continuation of the earth-shattering Palestinian victories of that day would have boosted the axis extending from the resistance in Gaza to that in Lebanon and Yemen and Iraq, and further to Iran, and further to Russia and China – a counter-alliance that now has an objective existence in the theatres; although, it should be noted, it is far looser, less coordinated, less dedicated and, of course, less powerful than the Western alliance.

Lastly, let me point out one more error of the lobby theory, perhaps the most damning one. It posits as the counterfactual a situation in which the US empire would be free to engage in rational deliberations and worry only about its own interests. Then it would ditch Israel, because how could it possibly stick up for something so destructive as the never-ending colonisation of Palestine, something that engenders such extensive and endless destruction in that land, in the region, beyond, all over the place. Surely this can't be what the US would opt for of its own volition? The error here is more than one, as it pertains to the natures of empire and capital and interest and rationality, but I will point to only one aspect. Considering how the US has consistently led the expansion of fossil fuel production and consumption around the world, after it took over leadership from the UK, and leads the speed-up of that expansion in the very moment when its destructiveness is plain to see and increasing by the day, it doesn't appear as much of a mystery that it also advances the destruction of one little land between the river and the sea. And no one, I think, could seriously argue that the reason we use fossil fuels is that the fossil fuel lobby in the US is powerful. It is, of course. But lobbies are surface phenomena. However mighty they may be, the fossil fuel and Zionist lobbies are epiphenomenal excrescences from deep structures that have operated over a very *longue durée*.

On the final page of *The Ethnic Cleansing of Palestine*, Pappe writes, prophetically:

> The Palestinians can never be part of the Zionist state and space and will continue to fight – and hopefully their struggle will be peaceful and successful. If not, it will be desperate and vengeful and, like a whirlwind, will suck all up in a huge perpetual sandstorm that will rage not only through the Arab and Muslim worlds, but also within Britain and the United States, the powers which, each in their turn, feed the tempest that threatens to ruin us all.[119]

We can now recognise this as more than an incidental metaphorical overlap, because climate breakdown is precisely the tempest that threatens to ruin us all, and the only thing the great powers have done to date is to feed it.

Destroy What Destroys Palestine and the Planet

Before I wrap this up, let me just propose some further moments of articulation in the present, in elliptical form.

The destruction of Palestine and the destruction of the earth play out in broad daylight. There is a surfeit of documentation for both. Knowledge of the two processes and how they unfold in real time is superabundant: we know everything we need to know about the catastrophes, and yet the capitalist core keeps rushing fuel to the fireplaces and bombs to Gaza.

Destruction and construction are interpenetrating opposites that presuppose one another: the destruction of the planet is the construction of fossil fuel infrastructure; the destruction of Palestine is the construction of racial colonies – or as Theodor Herzl put it in 1896: 'If I wish to substitute a new building for an old one, I must demolish before I construct.'[120] Limiting, stopping, reversing the destruction of Palestine and the planet therefore require, as a logically unassailable condition, the destruction of fossil fuel infrastructure and racial colonies – not necessarily their physical destruction; but necessarily their decommissioning and repurposing, in the cases where that is possible, and where not, on the path to their abolition, yes, their physical destruction.

It is entirely evident that investment in fossil fuel infrastructure must end and indeed should have ended long ago. Yet we see more pipelines, more rigs, more platforms and terminals and mines planned and built, and the more of them there are, the more difficult it becomes to cut emissions; the more fixed capital is sunk into the ground, the greater the imperative to maintain it and defend it against any transition away from fossil fuels. It is entirely evident that investment in racial colonies too must end and indeed should have ended long ago. Still, we see more settlements, ever more settlements planned and built on the West Bank and in Jerusalem, and perhaps soon in Gaza again. And the more Palestinian land is confiscated, the more housing units are erected and reserved for Jews only, the more difficult it becomes to envision a withdrawal to the Green Line, the more immovable the occupation, the greater the interest in defending it against any scheme for a viable Palestinian state.

This analogy at the level of the material base – creating ever more facts on the ground that prolong and intensify business as usual – is reflected at the level of superstructure. We keep hearing governments in the West talking about one and a half or two degrees and about a two-state solution, while actually existing processes of investment work ceaselessly to make both goals physically impossible. Talk of two degrees or two states here takes on the character of ideological cover. The parallelism is quite astonishing when one juxtaposes the COP summits with the summits of what was once known as 'the peace process'. Both commenced at the same moment in time, in the early 1990s; both had the function of sustaining the illusion that the so-called international community was working to mitigate climate change and giving the Palestinians their own state, respectively. Both operated with the same vacuous diplomatic rituals and incantations. Both covered up for the continued investment in destruction. But today, of course, only one of them remains: later this year we will have to suffer through the twenty-ninth edition of the COP circus, the next more vacated of meaning and substance than the one before it; there are no longer any handshakes outside the White House. 'The peace process' ended in 2005, when the state of Israel reconfigured its occupation of Gaza into the operation of a concentration camp. All that then remained was the naked, never-ending Nakba. Here, too, the catastrophe of Palestine appears to adumbrate that of climate.

A Lesson in Bourgeois Coldness

The genocide in Gaza provides an object lesson in callousness. In the climate catastrophe, the lives of non-white multitudes in the Global South do not count. They are expendable, of no value. We saw this play out in the disaster that struck Derna: more than 11,000 people killed over one single night left only the tiniest trace in the media of the West and none whatsoever in its politics. Just imagine if these had been 11,000 white Americans or Brits or Swedes killed over one night – imagine if it had been 11,000 of the people that really count: imagine the uproar. But here it was just the wretched of the earth, dying as they always do, in the Mediterranean and other graveyards of the world, their deaths part of the natural order of things, no notice paid to the fact that the excess of carbon in the atmosphere that killed them was put there by the rich people of the Global North. Instead, if there was any talk of blame and culpability in the Western media, the Libyans themselves were held responsible: had they not built such weak dams on that river, Derna would have withstood the pressure.

In the land of Palestine, the lives of Palestinians do not count. They are completely expendable. They have no value, none at all. This is the lesson we have learnt, once again, in the past half year – never has it been demonstrated with such extreme cruelty and indeed exterminationist bloodlust as now. Just imagine if it had been 40,000 Americans or Swedes or, most obviously, Israeli Jews killed in this manner – no, I think this is not something that can be imagined. It defies the political imagination. It goes beyond anything that could happen inside the world as we know it. And then the death of the Palestinians is also their own fault, highlighted with particular insistence when the genocide took off: the mass killing happened because the Palestinians send their own rockets crashing into hospitals; because they use their civilians as human shields; because they place their weapons in or next to schools and residential buildings; because of what they did on 7 October.

The genocide then curves back on the warming world and reconfirms the expendability and valuelessness of non-white lives: another *sine qua non* for the continuation of business as usual. It is very good for ExxonMobil and BP that the US and the UK have decided that death of this kind is *de rigueur.* The advanced late capitalist genocide reproduces ammunition for the paupericide.

The First Technogenocide

Much more could be said – and, thankfully, much superb work is now being produced – about the political ecology of the settler-colonial project in Palestine and the destructive tendencies that inhere in Zionism.[121] In Gaza, where it has been going on for decades, that destruction has now reached apocalyptic proportions: the people who have not yet died from the bombs live in a wasteland of undrinkable water, unexploded munitions, untreated sewage, overflowing landfills, contaminated soil, toxic debris, orchards and fields reduced to dust in this hyper-polluted strip of land in which human life is being rendered impossible for the long term.[122] Ecocide here fuses with genocide in a manner never seen before. Bosnia was not a less habitable land after 1995 than before 1992. Rwandan soil and water and air went relatively unscathed through the slaughter of hundreds of thousands of Tutsis. But will people ever be able to live again in Gaza? This dimension of the ongoing genocide blends with another, which has to do with the nature of the events on the morning of 7 October.

For empire and entity alike, the most shocking part of Tufan al-Aqsa was the way the resistance negated all the technological domination over Palestine in one fell swoop. All the walls of hardware built up over two centuries came down in a few hours. The *Jerusalem Post* composed a lamentation:

> How did an armed terrorist group succeed in overcoming the defenses of one of the most powerful militaries in the world? This is a question that will be asked for a long time [. . .] The epic shock from this attack raises questions about Israel's ability to

> confront other enemies. On the border on October 6, there was all the best technology. There were observation towers and soldiers observing Gaza. Israel also has drones and observation balloons [but] all the smart technology Israel has was rendered almost useless by the massive attack.[123]

Or in the words of two experts in the Global Network on Extremism and Technology:

> Home to leading military and defence engineering programmes, Israel watched its multimillion-dollar defence system struggle against forms of low-tech warfare [. . .] The 7 October attacks show that technologically inferior actors remain highly capable and dextrous against better-equipped state adversaries [. . .] High-tech defence means everything and nothing.[124]

The importance of the instant and complete negation of technological superiority on the morning of 7 October can hardly be overstated. It has no precedent in the history of Palestine. There is, of course, a history of guerrilla struggle, going back to the days of Izz al-Din al-Qassam, inflicting minor defeats on the enemy every now and then. The resistance has always been aware of the asymmetry: as the PFLP writes in its *Strategy* document, 'one of the enemy's basic points of strength is its scientific and technological superiority, and this superiority is reflected strongly in its military capabilities which we will face in our revolutionary war. How can we face and overcome this superiority?'[125] Tufan al-Aqsa provided the most resounding answer to that question ever registered: never before had the resistance swept away the accumulated technological forces of empire and entity with such supreme celerity and facility and comprehensiveness, the asymmetry turned upside down along an entire section of southern Palestine. No Palestinian uprising had accomplished anything close to this. A common comparison is with the surprise strikes of the October war in 1973, but they were delivered by the standing armies of Arab states. When it set out from the refugee camps in Gaza on the morning of 7 October, the Palestinian resistance struck from a position of seemingly absolute technological inferiority – although, granted, some of that inferiority had been ameliorated since the first intifada took off from the refugee camps of Gaza in December 1987. Back then, the Palestinians had only stones and at most a few knives; now they had rockets and RPGs and rifles and a handful of drones and the unforgettable paragliders – but still, nothing compared to the army they took on.[126] For the first time, the formula in place since 1840 was ripped apart: Palestinians themselves smashed through the technological apparatus dominating and destroying them.

One searches in vain for a similarly sharp inversion of a similarly wide asymmetry in the annals of anticolonial insurgency. The Tet Offensive has been invoked; but the Vietcong was a military force far better equipped than the Palestinian resistance. Guerrilla groups from

Cuba to Kenya overwhelmed adversaries with superior resources; but their superiority was never anything like Israel's on 6 October. The great affront of Tufan al-Aqsa was to shatter the complex of qualitatively superior military technology built up over two centuries: and because this must not be allowed, the punishment would have to be limitless. Those who think Israel would have responded less ferociously had everyone who died on 7 October held a gun fool themselves about the nature of that state. The simplest proof is what happened here in 2006: Israel resolved to destroy Lebanon after three of its soldiers had been killed and two abducted. So, what would it do after the scenes on the morning of 7 October? But the blow was hard not only to Israel. The US could not accept that the resistance flashed through its primary base in the Middle East as if it were a spider's web; it could not afford to see its own military machinery so humiliated. Israel and the US shared the imperative of restored deterrence.

What they have done together since 7 October has an easily decoded meaning: once we have repulsed the first blow, we will roll out all the forces of destruction we have in our stores. After the initial rout, we must rehabilitate our technology by reactivating its full capacity for annihilating life. The only way to undo the negation is to extravagantly reassert our full-spectrum dominance. This message is broadcast far beyond the borders of Palestine. It says: if you dare to pierce our armour like the Palestinian resistance did on 7 October, we will obliterate you and your people. The message is communicated not least to Lebanon; much as Charles Napier threatened to turn Alexandria into Akka, Yoav Gallant has repeated that 'what we did in Gaza can also be done in Beirut'.[127] But at stake here is the position of the US empire and its allies wherever it might face some kind of subversion. This war has a performative element in its display of technological superiority, a disinhibited flaunting of prowess – hence the videos where soldiers gloat over the detonation of family homes or schools.

Perhaps we can then specify this as the first techno-genocide. A technogenocide would be defined as a genocide that is 1) executed by means of the most advanced military technology, and 2) at least partly animated by the drive to restore its supremacy after a humiliatingly successful challenge. The genocide of the Bosnian Muslims was largely carried out with handguns, which the Sarajevo republic owned as well, albeit too few. The genocide in Rwanda was mostly effectuated by machetes. The Daesh genocide against the Yazidi was another low-tech genocide; while the paradigmatic case of a high-tech genocide, the Shoah itself, was never in any way provoked by a Jewish sapping of German technological power. Only the ongoing genocide in Gaza seems to fulfil both criteria. The Palestinians often refer to the 'Israeli killing machine', and that is precisely what it is: a machine for killing people, partly for the sake of rearming the reputation of the machine itself. The mass killing is mechanised and automated, as we know since the first revelations about the AI system called 'the Gospel' that processes enormous amounts of data about the civilian population and infrastructure to generate so-called power targets for the occupation army – 'a "mass assassination factory," in which the "emphasis is on quantity and not on quality."' Sources from within the army said: 'It really is like a factory. We work quickly and there is no time to delve deep into the target. The view is that we are judged according to how many targets we manage to generate.'[128] This is the killing machine in action, combining the muscle of petroleum with the mind of algorithms. Then there were the second revelations about the AI systems 'Lavender' and 'Where's Daddy?' that mass-produce kill lists with any number of civilians attached: as if the occupation decided to kill without inhibition and delegated oversight to the killing machine itself.[129] Because high-tech supremacy came to mean nothing on the morning of 7 October, it had to become everything again.

The Resistance Stands

But the Palestinian resistance still stands. After half a year, the resistance is still struggling. After half a year, six months, 184 days, the resistance is still fighting back on all fronts, from Beit Hanoun to Rafah and, of course, beyond Gaza. After all this time, Izz al-Din al-Qassam and Mohammed Deif and Abu Obeida and their comrades-in-arms from Jihad and the DFLP and the PFLP are still in the tunnels, still dispatching one operation after another – and this is what makes it possible to live another day. I work in the West, in academia, in the production of knowledge and ideas. An absurd situation prevails there. It is possible to ignore or condone or justify or praise the genocidal politics of Israel, without risking anything, without being disqualified from anything or losing any respectability. But supporting the Palestinian resistance – the armed resistance, the only force opposing the genocide on the ground – is prohibited. I, for one, refuse to go along with this. I think the real disgrace in the West is that the left cannot clearly and without equivocation support the Palestinian struggle for self-emancipation. This is a topic for another lecture and many texts, but I think we should say it loud: we stand with the resistance and we are proud.

Response to Some Objections Regarding the Palestinian Resistance

The publication of the above text provoked some rejoinders, none concerned with the events of 1840 or the other historical material. On the Verso Blog, Matan Kaminer wrote 'After the Flood', a comradely critique which I much appreciated.[1] He argues that the leadership of the Palestinian resistance, Hamas in particular, is not necessarily a partner 'in the struggle for a shared and inhabitable Earth'. I responded in a text written in late May 2024, reproduced below with mostly minor changes and a few substantial additions.

On the Left Side of the Resistance

Matan Kaminer has no quibbles with the historical analysis presented here, but a few with my support for the Palestinian resistance, which, admittedly, was only roughly sketched. In the spirit of comradely critique, I shall here respond to his objections, and take the opportunity to elaborate a few points.[2] But let me begin with some first principles and personal points of departure.

If people identify as Marxists, what ought they do if they were in Gaza? They ought to join Marxists fighting on the ground there. And if they are unable to enter the besieged and destroyed and occupied and burning ghetto? Then they should stand in solidarity with their comrades inside. These are not entirely hypothetical matters. It so happens that there *are* Marxists fighting on the ground, namely the Popular Front for the Liberation of Palestine (PFLP) and the Democratic Front for the Liberation of Palestine (DFLP). From the earliest morning hours of Tufan al-Aqsa to the battles raging yesterday, they have participated in the guerrilla war against the genocidal occupation. They have conquered towers, taken prisoners, fired rockets, laid ambushes, shot down drones, blown up tanks, confronted soldiers in close combat in the alleys of the refugee camps and retreated into the tunnels and persevered throughout all these months of unfathomable horror. They have lost fighters, cadres, elected representatives in action and in the targeted bombings of civilian homes.

The Abu Ali Mustafa Brigades of the PFLP have been active from Mukhayyam Jenin to Mukhayyam Jabaliya, Ramallah to Rafah. But a remarkable aspect of Tufan al-Aqsa has been the rise of the Omar al-Qassem Forces of the DFLP: we don't have reliable statistics, but the stream of resistance operation reports suggests that they have vied for third place among the armed forces in Gaza. If these two Marxist guerrilla forces are grouped together – and if the al-Asifah Forces and the Jihad Jibril Brigades are added to the mix – the left bloc would make up the second largest on the ground, after the bloc consisting of the Izz al-Din Qassam Brigades of Hamas and the al-Quds Brigades of Jihad. And – of course, self-evidently, naturally – all of these factions fight together. They fire rockets in shared operations and coordinate manoeuvres and slip back into the same underground bases. This is as it should be. There is a time for sharp debates about what a liberated Palestine should look like. There is a time for unity in action. If ever there was a time for the latter, that time is now: and the resistance has indeed been perfectly united from day one into the ongoing eighth month.

How shall we assess the presence of the left here? Kaminer laments 'the fact that the resistance is not being led by a secular, democratic force such as the PFLP'. This is true, of course. The resistance has never been led by the PFLP or the DFLP; these have always been the left currents of the movement. Granted, the PFLP is, for a number of reasons, far weaker today than it was during the golden age of armed struggle in the diaspora, 1968–73, and the heydays of the first intifada, 1988–90. But even at these points, Fateh was the dominant force. Today it is Hamas and Jihad. The weakness of the left compared to its older peaks can hardly be a reason not to stand in solidarity with its struggle. 'No reader of the Verso Blog needs yet another reminder of the cataclysmic defeat the left has suffered across the globe since the 1970s; but there it is, yet again, in all its depressing ugliness,' Kaminer writes. Well, readers are encouraged to go online and watch a video that gives a glimpse of the strength and beauty of the PFLP in Gaza one year before the Tufan al-Aqsa operation.[3]

Precisely the dismal status of the left across the globe should make us value this one all the more. Here we have it, an actual organised left, present on the frontlines of the central battle against empire in this historic moment. One could compare this with the fate of the left in Iran or Egypt: neither the Fedaiyan-e Khalq nor the Revolutionary Socialists exist any longer (and not because the tyrannies they fought have vanished). That should be further reason to appreciate this exception to the rule. But the left in the Global North has followed events since 7 October by paying little if any attention to the left on the ground in Palestine – as if it forgot about the PFLP and the DFLP once the Oslo Accords were signed, or became so used to the idea that progressive forces no longer wage armed struggle that it doesn't care about those who do. I find this annoying. The least any honourable Marxist can do at home is to follow the steady stream of communiques and analyses from the two Fronts, through, for instance, the Resistance News Network, the invaluable Telegram channel – to pick just one, here is Jamil Mezher, deputy general secretary of the PFLP, on May 12:

> The Palestinian resistance in Gaza, with its military wings, is engaging in the greatest epic of this modern era. Despite its modest means, its joint operations in the field and surprise tactics have managed to defeat the strongest military force in the region, heavily armed with all types of American-made weapons, and permanently undermine its deterrence and military superiority. The Zionist entity has failed in marketing its colonial project and its claim as the only democratic state in the East, revealing its racist identity and ugly face to the entire world during the eight months of the Al-Aqsa Flood battle. We witnessed the victory of the Palestinian right, the historical narrative, and the legitimacy of our resistance.

Don't fall for the luxury of ignoring or sneering at this left. I advise comrades to tune in to it – not as part of a 'rushing to proclaim that we "stand with the resistance"' (Kaminer), but as part of a commitment forged long ago and enduring until the day of victory or defeat. This is a matter of *muqawama wa thawra haya al-tahrir wa al-awda*, as in the headline of the latest edition of the *New York War Crimes*, the newspaper produced by the Palestine solidarity movement in this city: 'generation after generation until total liberation' – no rushing, no matter of fashion: a covenant for the ages.

On a Personal Note

Allow me to insert a personal note here. Since I first travelled and lived in occupied Palestine in the late 1990s and established contact with it, the PFLP has always been my favourite faction. For me, solidarity with the Palestinian resistance in general and its left in particular began long before 7 October and will continue long after; it is among my three or four deepest political convictions. I wrote my first book (in Swedish) in 2002, based on my experiences of April that year. An American comrade and I happened to be the first outsiders to enter Mukhayyam Jenin while the battle was still raging. We moved among the decomposed corpses, the pulverised buildings, the refugees who had their camp destroyed over their heads: the Palestinian ur-scene, eternally extrapolated on ever-larger scales, today in Jenin again – this endlessly tortured camp – but to an incomparably greater extent in Gaza. We tried to do what little we could to get some aid in. We also stumbled upon the severely wounded third leader of Jihad in his hiding place and brought him out for medical treatment. Sometimes I think that is the single most meaningful act I have performed in the realm of politics: obviously the life of a Jihad leader in Mukhayyam Jenin is worth more than 1,000 texts. This is where I come from; this is who I am; climate, for me, came later. One reason I was so struck by that problem was that it appeared to me as a planetary version of the Nakba. Here it is not one homeland that is being destroyed, but an entire liveable planet. It was something of this fractal pattern I sought to capture in my lecture.

So, personally, I came to climate through Palestine, and if I have sometimes called for the climate struggle to get real and militant, it is because Palestinians taught me the meaning of resistance. This emphatically does not mean the climate movement should copy the tactics of the armed struggle – firing rockets or sniping soldiers or any other kind of targeting of human bodies – for a host of manifest reasons, one of them being that fossil capital is not a settler-colonial project. But there is one thing that should be emulated, and that is the *spirit* of resistance, the readiness to fight as though every part of one's existential dignity hinges on it. If that spirit can be mobilised for a homeland, it should, one would think, also be possible to have something like it for the earth. In any case, I have come to think that the meaning of life is to never give up – no matter if it is too late to prevent catastrophe; no matter how many disasters pile up; no matter how overwhelmingly powerful the enemy. And there is no force in the world today that embodies this meaning like the Palestinian resistance.

If I lived in Gaza, I would, I imagine, be a long-standing member of the PFLP. (If I were a woman, I would have joined the women's brigades of the DFLP.)[4] I imagine that if Marxists were able to enter Gaza, some would have gone there and enlisted with either Front, much like some went to Rojava to fight Daesh. Again, this is not an entirely hypothetical matter. I know of comrades who have tried to make it into Gaza, in the best tradition of internationalism, to join the struggle. Suffice it to say that the way the Sisi regime collaborates in the destruction of Gaza makes it impossible for volunteers to enter. Had these comrades succeeded, they would, of course, have stood shoulder to shoulder with fighters from al-Qassam and al-Quds and accepted their tactical leadership on the battlefield. What could be the objections to that?

Palestine with or without Fossil Fuels

Kaminer takes the climate connection as seriously as it should be taken and retorts that the resistance is itself soaked in fossil fuels, if only indirectly, via Iran and Qatar. He is right, in a manner of speaking. The ruling class of Iran is utterly based on profits from constantly expanded oil and gas extraction (I have written about this in one book and one article).[5] Qatar presents the same picture. Should we therefore part with the resistance on climate grounds? Marxists tend to be fond of historical analogies, and I too shall fall for the temptation: consider the case of the Holocaust. It was a genocide executed by means of coal, by a Nazi regime in love with every fossil-fuelled machine of domination, as we argue in *White Skin, Black Fuel: On the Danger of Fossil Fascism*. The destruction of the European Jews was also a moment in the destruction of the earth.

Complicating things, however, is the circumstance that the Third Reich was crushed by the Soviet Union, partly because the Stalinist regime had so much more oil and fossil-fuelled machinery at its disposal. It also deployed and supported partisans across Europe during the war. Here, fossil fuels were mobilised for survival and liberation. This does not diminish the radiative forcing of the allied emissions, nor the severity of the postwar boom in fossil fuel production in both East and West, but I don't think we can really regret this mobilisation.

Put it this way: because all modern economies have been totally predicated on fossil fuels, even a few good things that have happened in modern history have been sullied with them. We should be thankful for the Stalinist regime using its reserves the way it did between 1941 and 1945. We should also be grateful for the support it gave, from its fossil-fuelled pockets, to the National Liberation Front (NLF) and the African National Congress (ANC) and the PFLP and the other anti-colonial liberation movements of the Cold War era, even as it maintained its suffocating and sclerotic tyranny at home. *Mutatis mutandis*, I don't see how we could wish for the funding of the resistance to be severed. I would rather have oil profits from the Middle East being recycled into it or into Al Jazeera – that single source of sustained sanity in the global media landscape (how would we live without it?) – than into the IDF or the Sisi regime.

Which is not to say, of course, that fossil fuel production in Iran or Qatar can go on longer than anywhere else. All of it must stop. If a Palestinian state were to ever become a reality, extraction of the gas in its waters would have to end too. But, yes, as Kaminer points out, there is no reason to believe that Hamas – or the PFLP or the DFLP or any other Palestinian faction – would, in its current iteration, elect to forego that source of revenue. Before the bombing began in October, Gaza was probably the part of the world most fully dependent on solar power, but this was a function of its being denied access to other sources of electricity; any Gazan would jump at the opportunity to burn gas or coal. The government of the Strip has no visible climate policy.

This is part of the tragedy: it is easy to demonstrate that the destruction of the earth is bound up with various localised forms of destruction, but much harder to point to movements that rise up against the former. The Palestinian resistance is evidently not a climate movement. Similarly, one could show that the destruction of Yemen is part of the generalised biospheric process, but there is no discernible agent of climate rebellion in that country. We are grappling with a structural deficit of climate subjectivity and a structural surplus of objective forces of destruction; and perhaps the imbalance is nowhere as extreme as in the Middle East. (Latin America is far richer on the subjective side.) Those of us who care for or work on or live in this region have an enormous task ahead of us. If some climate awareness might have been budding in Palestine over the past few years, part of the tragedy is, I suspect, that the ongoing genocide will drive it down on the agenda. But I fail to see why any of this should prompt us to distance ourselves from the resistance. Injecting climate consciousness into Palestinian and other struggles in the Middle East is a necessary historical mission, but it cannot deflect from the immediate defence of the existence and rights of the Palestinian people.

On Killing Settlers

Two objections remain regarding the events on 7 October and the nature of Hamas. On the first, we were recently blessed with a brilliant analysis from Adi Callai, an Israeli comrade.[6] We have seen the fabricated stories about what happened methodically debunked.[7] We could go on discussing the details of what took place for years, and we probably will, but let us here, for reasons of length if no other, stay with the central question: should we condemn or renounce Tufan al-Aqsa, or the resistance as such, because it crossed the line of killing civilians on that day? One can adopt such a principled position. Or one can – staying with the method of historical analogy – consider the first car bomb planted by the uMkhonto we Sizwe (MK) in 1983, aimed at an air force and military intelligence office in the heart of Pretoria, in which nineteen people were killed, white civilians included. After this bomb, the armed wing of the ANC entered a phase of escalated struggle against the apartheid regime, causing a string of civilian deaths. What should we think of this?

> The killing of civilians was a tragic accident, and I felt a profound horror at the death toll. But as disturbed as I was by these casualties, I knew that such accidents were the inevitable consequence of the decision to embark on a military struggle [. . .] As Oliver [Tambo] said at the time of the bombing, the armed struggle was imposed on us by the violence of the apartheid regime.[8]

This is Nelson Mandela in his autobiography *Long Walk to Freedom*. One can, of course, slam him as far too extreme on this score – a terrorist. The US government designated him as such until 2008.[9] People in the Global North tend to be scandalised by the realities of the struggle against colonial regimes exported from their latitudes; people in the Global South, less so. Nelson Mandela was the person who would solemnly raise his fist and hum along with his comrades in the classic MK song with the refrain 'We the members of the MK have pledged ourselves to kill them, the whites.'[10] He was the person who visited Gaza in October 1999 and said to an audience of Palestinian politicians: 'Choose peace, rather than confrontation – except in cases where we cannot proceed, where we cannot move forward. Then, if the only alternative is violence, we will use violence.'[11] My hunch is that if this person were alive today, he would know whom to support.

But yes: tragic losses, deplorable, nothing to gloat over, nothing to celebrate or take pride in. Here Hamas's attitude to the civilian deaths it caused on 7 October is not so different from Mandela's.[12] What it is justifiably proud of are the losses it inflicted, and continues to inflict, on the occupation army. Personally, I too felt horror at some of the things Palestinians did on that day. I also don't particularly enjoy reading about the killing of civilians at the hands of the FLN in Algeria or during the Mau Mau rebellion, the Nat Turner revolt, the Haitian Revolution, the Sepoy mutiny or the Tupac Amaru uprising, to mention just a few cases – several of which resulted in gruesome mass violence against the civilian population of colonisers to an extent far beyond anything Palestinians have ever done. And yet we, on the left, commemorate the Haitian Revolution as the greatest single act of emancipation in the New World, perhaps in all of the modern era. Should we, after 7 October, start condemning it on the principle that civilians must never be killed? Or should we remember the violence against the white civilians as an ugly aspect of a legitimate struggle for freedom, the victory of which marked an unusually genuine instance of progress in history?

These are difficult matters, I concede. What, for example, could Mandela possibly have meant when he wrote that losses of civilian life were 'the inevitable consequence of the decision to embark on a military struggle'? The NLF didn't kill civilian Americans in Vietnam. But then again, the American occupation of Vietnam was not a settler-colonial project. The 'inevitability' Mandela refers to would seem to result from the nature of settler colonialism: when colonial regimes from the North implant settlers among the indigenous and/or imported slave population to displace and/or exploit and/or exterminate it, attempts by the latter to shake off the yoke and attain minimal freedom and ensure survival come up against their presence. During the Second World War, the partisans in Norway had only soldiers to fight against. But had the Third Reich placed, say, 100,000 Aryan settlers in their midst to torment and destroy them, I bet even the meekest Norwegians would have, at some point, crossed the line. This does not resolve all the problems of what happened on 7 October. It does, however, in my view, call into doubt the position that anti-colonial struggle must be disavowed the moment civilians from the settler population fall. If violence against settlers – including those who do not hold a gun in the moment of attack – is an inevitable component of the struggle against settler-colonial oppression and/or annihilation, as every case from the Americas to Algeria and beyond seems to suggest, then the boot is on the other foot. What struggle in this category could those who renounce or denounce a resistance for crossing the line still support? Where would they end up in history, with such a compass? Just mentioning the name of Fanon should settle this question; but it is not a matter of retrospective solidarity with the coloniser or the colonised. It is now a matter of the most urgent present.

Choose peace, except in cases where we cannot proceed, where we cannot move forward, said Mandela of the MK: did the situation in Gaza before October 2023 meet this criterion? I cannot see how one could possibly come to any other conclusion than it had been met several times over. The Palestinian national movement – Hamas very much included (but not the collaborationist *sulta*, of course; it has long ago left that movement) – had tried every conceivable route to gain a minimum of freedom for its people. This included peace offers, negotiation initiatives, boycott campaigns, appeals to international institutions, non-violent marches and mass demonstrations and civil disobedience culminating – as should never be forgotten – in the Great March of Return of 2018–19, during which the occupation army murdered 223 Palestinians and maimed or otherwise injured some 10,000. If ever armed struggle has been imposed on anyone, this is it. Much, much more could be said on these issues, but I have yet to see anything that could overturn the status of Tufan al-Aqsa – from 7 October to the most recent battle – as the greatest anti-colonial revolt of the twenty-first century.

On Actually Existing Hamas

As for Hamas, we could and should have as long a conversation about what that movement really is, how it has developed, what it stands for, where it is going. That would begin with abandoning the liberal discourse about Hamas, because it does not engage with reality.[13] It is a psychic phenomenon fuelled by fears and fantasies about demons. Serious study of Hamas benefits from scientific work on the movement by scholars like

Sara Roy or Tareq Baconi or – his being one of the finest monographs on the topic in recent years – Somdeep Sen.[14] The same applies to Jihad.[15] Apart from its links to the fossil-fuelled Persian Gulf, Kaminer levels three charges against Hamas: it is guilty of 'messianism, authoritarianism, and sectarian manipulations'. Let us deal briefly with each.

Messianism is not a thing among Hamas or any other part of the Palestinian resistance. It is a thing among the activists of the Temple Movement, who are preparing to slaughter their red heifers, raze al-Aqsa, build the temple and inaugurate the Messianic era, and who, along with their fellow religious fanatics, exercise ever more power over the entity, as even the *New York Times* recently felt compelled to chronicle.[16] But Hamas has no messianic notions. Contrariwise, the trajectory of its almost four decades of existence is, as the aforementioned scholars and anyone else who has followed the movement can attest, one of steady secularisation. The features of that process have accumulated persistently: the break with the deranged and inexcusable anti-Semitism of the first charter, the loss of interest in gender segregation and hijab imposition, the divorce from the Muslim Brotherhood, the de-emphasising of piety and focus on politics, the framing of that politics in less Islamic and more strictly national terms. Hamas today and the movement in early 1988 – not to mention the Mujama al-Islamiya from which it sprang – are two different species. One can familiarise oneself with actually existing Hamas by following, as I think everyone should, the speeches given by Abu Obeida. Since 7 October, they have usually opened with the Islamic salutations and a token quotation from the Qur'an – some sura about how the weak will ultimately defeat the strong or about freedom finally coming – before quickly veering into expositions on the challenges facing the resistance, the achievements, the sacrifices, the crimes of the occupation, the virtues of solidarity, the path ahead. Don't be scared! It's great stuff, listen in. The messaging is thin on religion but laden with anti-colonial, anti-fascist rhetoric, and the same goes for Hamas's daily communiques.

But the secularisation is perhaps most remarkable at the level of military doctrine. After the massacre in the Ibrahimi Mosque in 1994, the early generations of al-Qassam developed the tactic of martyrdom operations, also known as suicide bombings. They were central to the second intifada and drew on – among many other sources – the religious notion of martyrdom as a value in and of itself.[17] But during this war, there has not – contrary to the expectations of some – been any report of *a single martyrdom operation* in Gaza: not one suicide bomber, nor any other kind of action in which the fighter intended to die.[18] With occupation soldiers spread out in hospitals and apartments and checkpoints, the opportunities have been ample. But Muhammed Deif and the other leaders of al-Qassam have discarded the old tactic. The method is for the fighters to emerge from the tunnels and place their bombs under the Merkava tanks or whatever other target they choose and then run and climb back down to safety. This is guerrilla warfare on the classic model (but with a tunnel system instead of a rainforest), secular in form, often refined to tactical perfection.

Hamas is, of course, far from consummately secular; it is still very much the Islamic resistance movement. But its struggle has shifted radically from a religious to a political register: a de facto secularisation all the more noteworthy for the exponential rise of religious Zionism, which has never been more extreme than during this genocide. As I write, the occupation army has just blown up the main water reservoir serving Rafah. A soldier bragged about the act on social media with the following words: 'The destruction of Tel Sultan's water reservoir in honour of Shabbat.'[19] Such religiously justified depravity has become ever more common on the side of occupation. Next to the social media accounts of its soldiers, those of Hamas look like wonders of rationality and enlightenment. I hope the communiques from the entire period of Tufan al-Aqsa will soon be collected in one volume so readers can judge for themselves.

What of the authoritarianism? Hamas is no absolute paragon of democratic practice. But surely things are relative. If we adopt the Marxist view of bourgeois democracy and treasure it for the room it gives to the revolutionary left to operate freely, we must conclude that Gaza under the rule of Hamas was the most complete instantiation of such democracy anywhere between Beirut and Tunis. The PFLP and the DFLP could organise mass rallies and training exercises for separatist women's brigades and any other activities with a freedom found nowhere else in that zone. Hamas did not murder Nizar Banat.[20] Hamas did not torture Basil al-Araj.[21] The restrictions it imposed on the population paled in comparison to the vicious repression of dissent and struggle in the enclaves controlled by the *sulta*, the Palestinian Authority, whose sole *raison d'être* is collaboration with the occupation. It is against this background that the current situation has developed: relations between the Marxist and Islamic blocs of the resistance have never been better. Not only did Tufan al-Aqsa set out from a joint operations room, but the leadership of Hamas has consulted with that of the PFLP (less so with the DFLP) along every step of this painful way, concerning everything from demands in the negotiations to approaches to external actors.[22] If it has been brutally authoritarian to anyone over the years, it is Daesh. Gaza stood out for the total inability of takfiri Salafi jihadists to gain a foothold there, because when they tried, Hamas would smash them. Again, this is not to say that Gaza under Hamas was some sort of democratic paradise – how could it possibly be, under the conditions of the siege? – but an organisation more prone to authoritarianism could easily have created a full-blown despotic dystopia. The police state of Sisi and the police non-state of Abbas are proof enough.

This brings us, finally, to the charge of sectarian manipulations. I honestly do not know what this could refer to. Insofar as there are any sectarian schisms within the Palestinian people, it would be between the Muslim majority and the Christian minority; but relations between Hamas and the Christian community in Gaza have been excellent, as even *Haaretz* recently acknowledged (although much more could, again, be said about it).[23] As for the great divide between Sunni and Shia that has so plagued the Middle East in the past decades, the consensus is that Tufan al-Aqsa has done more to heal it than any other event in recent memory – particularly in Lebanon, a country chronically vulnerable to it. If Kaminer has in mind manipulations between political factions, Hamas is second only to Jihad in efforts to secure national unity: it is, of course, the Fateh leadership that has spoiled every attempt at reconciliation and the restructuring of the PLO. Kaminer proposes Marwan Barghouti as a possible alternative pole. I agree: nothing would be more delightful than to have him out of prison. But the irony here is that relations between Marwan and Hamas have, again, become ever closer over the past years – this is no secret – as most evident in the long-standing top demand of Hamas in every negotiation with the occupation: let him go. This demand precedes Tufan al-Aqsa (and it also includes Ahmed Sa'adat).[24] When that operation was launched, a primary tactical goal was to capture prisoners, with whom the resistance could bargain and force the occupation to open its prison gates – above everyone else, for Marwan. He is still languishing in jail. But if there ever was a chance to get him out, Tufan al-Aqsa was – or is, insofar as the Netanyahu government still cares at all for its prisoners – it. We can count on him being aware of this. Let us hope that he will be the Palestinian Mandela; and let us not nurture any illusions about him emerging from his cell as an enemy of Hamas.

Stay with the Vow

Over the past month, I have immersed myself in the movement for Palestine in New York. There are many things that could be said about it and how great it has been, and here is one: I have never, in all my decades in this movement in the Global North, seen such pronounced support for the resistance. The silhouette of Abu Obeida hovered over the CUNY encampment. The red triangle is ubiquitous. At the demonstration in Brooklyn violently attacked by the NYPD on 18 May, young women without hijabs marched with al-Qassam and PFLP pins. Signs and banners included pictures of Abu Obeida, Sinwar and Deif alongside Sa'adat; 'resistance is justified when people are colonized' – a common chant; 'when injustice becomes normal, resistance becomes a duty'; 'power to all our martyrs – long live the Palestinian resistance'. You don't get as much of this in the streets of London and certainly not in Berlin. Should we deplore it? I, for one, consider it a splendid sign of radicalisation where it is needed most: in the heart of the empire. I hope this generation of activists keeps being inspired by the resistance to never give up and always press on, however oppressive the state it faces might be. On the same note, I think the most decent approach to the resistance has been that sustained on a daily basis by Nora Barrows-Friedman, Ali Abunimah and the other comrades of the *Electronic Intifada* – not the least Jon Elmer, who has had the best-informed commentary running day after day.[25] No one should miss it.

This is how things stand in the present moment; they could of course change at any time. Comrades should not only be critical towards each other but also open to revisions. One cannot exclude a scenario where the resistance in Gaza is fully drenched in blood and whatever fragments of it come out on the other side degenerate into some insanity: a revival of the 'external operations' (something the PFLP and the DFLP used to practise, but Hamas and Jihad have never done); a turn towards some more sinister version of Islamism. Then the solidarity with the resistance would have to be reassessed. Kaminer is right that we cannot stick to some transhistorical 'blanket approval for anything that any Palestinian does in the name of liberation' (I never quite liked Abu Nidal). But, for now, given the historical gravity of the genocide, and given the nature of the desperate resistance waged in Gaza – not to forget the northern West Bank – I think the situation is rather the opposite.

When we speak of resistance against murderous oppression and the fight for survival and freedom, do we only bother about the past? Are we happy with it when it is so comfortably distant as to be amenable to nostalgia? We have the Black Panthers in photo books and

Malcolm X on our walls – then why not also the PFLP and Abu Obeida? Is revolutionary politics a posturing about the past, or about real struggles that happen as we speak? On what grounds do we admire the heroes from millennia of subaltern endeavours at self-emancipation, but not the Palestinian fighters who run all the way up to the tanks and deliver their bombs with their hands and dash off? Why should they not be in our pantheon? I see two possible reasons: we are actually not that serious about the commitment to the struggle for freedom, or we do not consider Palestinian lives worthy of being fought for. If these reasons are excluded, the fighters who dare to blow up the Merkavas from zero distance ought to be regarded as the nonpareil heroes on Earth today. This fight may well end in defeat, in the total destruction of Gaza and its people, and more beyond. Until then – so the streets of New York tell me – there will be quite a few of us who stay true to the vow and stand with the resistance.

On Blaming the Fighters

In *Jacobin*, Bashir Abu-Manneh voiced another objection to my blog post, in a piece presenting a case against the Palestinian resistance in general and Hamas in particular. He attributes to me the view that 'the Al-Aqsa Flood operation achieved more than the First Intifada because Palestinians managed to replace stones with military arms – ignoring that the intifada was the largest self-organized anti-colonial mass movement in

Palestinian history, and that it compelled Israel to make unprecedented political concessions'.[26] But I did not make a total political balance sheet for Tufan al-Aqsa, nor did I for the first intifada. All I said was that no Palestinian revolt has accomplished anything like the negation of US–Israeli military superiority that occurred on 7 October. The first intifada was my first love in the field of Palestinian politics. It took about five years from 8 December 1987 for its results to become visible, and as Callai has argued in his unsurpassed analysis of 'the Gaza ghetto uprising', we might likewise have to wait five years after 7 October to know how it panned out in the end. It would be far too soon to say now that it achieved more for the Palestinian people than any previous revolutionary cycle. It also cannot be ruled out. The Zionist entity doesn't exactly appear to be in vigorous health these days.

But what about the genocide? This is the more substantial objection from Abu-Manneh: 'Indeed, to argue that Hamas has managed to achieve more is to totally ignore that its military attack has triggered a huge genocide against the Palestinian people.' The gist of his case is that Hamas bears responsibility for the genocide and should not have launched Tufan al-Aqsa, because it precipitated this catastrophe. Given how Israel responded, 7 October was a mistake. By the same logic, the Paris Commune ought never to have happened, as it elicited the massacre of some 20,000 to 30,000 Communards and other Parisians. The Bolsheviks should have refrained from seizing power, since they triggered an orgy of pogroms and the emergence of fascism in Europe and ultimately its *Endlösung*. Why couldn't the Egyptians just bow to the rule of Hosni Mubarak? And so on. To thus distribute culpability for the extreme excesses of repression is to blame the victims, and the fighters – as if those who seek freedom are guilty of the frenzied attempts to keep domination in place.

The state of Israel could, of course, have chosen to act differently after the morning hours of 7 October. It could have reconquered the territory seized by the resistance. It could have regained all the lands up to the borders of the Gaza Strip and then entered negotiations about an exchange of prisoners: cancelling the military gains of the Palestinians and dealing with this enemy to get the hostages back. The fact that the state did not behave in such a manner but rather plunged headlong into genocide cannot be attributed to any Palestinian actor. The sole explanation for it is the pre-existing hatred of Gaza, as the repository of the Palestinian claim to Palestine; the unlimited hatred of the people concentrated there, undergoing a well-documented growth since 1948; and most fundamentally, the burning hatred of the Palestinian presence in Palestine, which inflamed the Zionist project from the beginning.[27] The gist of my lecture was the exact opposite of Abu-Manneh's: this genocide has a set of very deep sources indeed. The resistance is the resistance against it.

Response to Some Objections Regarding the Israel Lobby

The critique of the Israel lobby thesis laid out in 'The Destruction of Palestine is the Destruction of the Earth' elicited a response from Ed McNally in *Jacobin* magazine. McNally argued that Israel is a liability for the US and its empire and therefore US commitment to supporting Israel can only be explained through the power of the Israel lobby. My response to McNally is published for the first time below.

In *Jacobin*, Ed McNally published a piece called 'The Israel Lobby Matters', which took aim at this aspect of my argument. He complains about a superficial empiricism, on display in my quoting Joe Biden about the need to invent an Israel. One should not read 'the existing interests of the American empire off cherry-picked pronunciations from certain of its leaders'.[1] This is a somewhat peculiar objection, since McNally's own restatement of the theory is largely based on empiricist methodology: adducing cherry-picked quotations by various representatives of empire – from Kissinger to Clinton – as evidence for the omnipotence of the Israel lobby. My quoting Biden was meant to put the most recent formulations in the context of two centuries of actual ideological and material invention of Israel. Biden's words are of interest because they accurately express a history in motion since 1840. That history is vast and contains so many protagonists untroubled by any Israel lobby – from the original British Empire to West Germany, the main backer of Israel after 1948, in which no such lobby existed for obvious reasons – that Occam's razor makes the theory superfluous.[2]

But McNally keenly reasserts its core tenet: Israel is a liability for the US, the alliance a source of self-harm that must be foisted upon the US by an alien actor. The Zionist entity in general and its present deeds in particular do not align with American interests. Ergo, the lobby must keep Washington in a stranglehold and enforce continued support. US leaders might *think* that such support is in their own interest, but McNally knows that it is not. 'US leaders', he writes, 'are more than capable not only of making disastrous strategic miscalculations, but of clinging onto wrongheaded conceptions about the interests of the empire they superintend.' Recycling the analogy used by Ghada Karmi, he argues that the invasion of Iraq was 'a net-negative for American power. Here was a disastrous ideological crusade, based on self-defeating hubris about the world-making potential of shock-and-awe military interventions.'

Two contentions are here rolled into one. First, if an act turns out to be a net-negative for an empire, it cannot have been motivated by its true interests. Second, a policy that keeps being counterproductive and going against those interests must stem from the outside – the role played by the lobby. Let us begin with the first contention. The logic here is that only an act that generates long-term benefits for an empire can possibly have been in its interest; if something goes awry along the way, it must have been propelled by something else. So because the invasion of Afghanistan was indubitably a net-negative for the Soviet Union, it likewise cannot have originated in Soviet interests, but must have been forced upon the USSR by some exterior body insinuating itself into the higher echelons of the Kremlin, duping its leadership into doing something so self-destructive. Obviously, we would now have to rewrite the entire history of empire, from pharaonic Egypt onwards. Every case of overstretch or failure must be reinterpreted as the effect of some manipulative lobby. Perhaps it was Israeli all along? Or we are here simply dealing with a badly non-dialectical conception of interest. If applied to phenomena other than empire, we would, for instance, have to conclude that financialisation could never have been in the interest of capital, because it created financial instabilities highly generative of crises. Perhaps some survivor in 2124 will be able to argue that fossil fuels can never have been in the interest of the capitalist mode of production, because global heating eventually destroyed it too; so there must have been some lobby bamboozling that mode all along.

The logic of these contentions is a recipe for analytical meltdown, but it does compel us to clarify a central category, namely that of interest. What do we mean when we say that something is in the interest of an empire? I think we have in mind something like the following: a political project impelled by an internally generated dynamic, working over a long period of time but condensed in each conjuncture and expressed through the actions and thoughts of the dominant classes of the empire, so as to defend and extend its power and, in the last instance – insofar as we are dealing with capitalist empires, such as the US – advance the accumulation of capital, in particular capital based in the metropole, such as American capital. This definition would clearly be compatible with malfunctions and misfortunes. A project can be in the interest of an empire and still, in the fullness of time, yield mostly trouble for it. Is this what Israel has done for the US? Surely the jury is still out on the question. McNally seems very certain that Israel has done more harm than good for the power of the US empire, between 1967 and 2024; on what evidential grounds, I do not know.

Granted, McNally does tell us that the Middle East has lost its strategic significance. The supposed proof for this lies in a reference to an article by two realist scholars who claim that the US would be better off dumping the region. Leftists in love with the lobby theory tend to be enamoured with realists. For the rest of us, it rather looks as if the US remains pretty heavily invested in the Middle East; that all talk of pivoting away from and abandoning the region has so far proven premature; that the heightened tensions with Russia and China intensify inter-imperialist rivalry in this part of the world, with Iran as a nerve centre; that the US – whether the president is a Republican or a Democrat – seeks to clinch domination of it by unifying the Zionist entity with the most reactionary Gulf capital. But, yes, the empirical details of this constellation do, I repeat, need to be studied anew and updated. The old theories of the PFLP and the rest of the resistance only provide us with guideposts. These seem to me, however, far more promising and productive than anything from the lobby theory, which axiomatically rules out a convergence of interest.

To bolster this axiom, McNally alludes to a school of revisionist historiography in the study of the British empire, which has claimed that the impetus to imperial expansion never operated inside the metropole, but at its far-flung outposts: colonisers at the margins lured Britain into foreign adventures, when London in fact had no desire or interest to engage in them. This is not the place to debunk this fantastically flawed rewriting of British imperial history. The historical material presented above should be enough to spot just a few of its errors. But McNally wants to use this revisionist school as a prop for elevating the structure of the lobby theory into a general, even transhistorical model for how empires work. Tails really do wag the dog. And indeed, if Israel is not in the interest of the US, surely one could also make the case that the occupation of the West Bank is not in the interest of Israel. (Some liberal Zionists do.) The occupation therefore cannot have its roots in the Zionist project but must have been grafted onto Israel by a foreign lobby – the settlement lobby, or perhaps an Israel-Israel lobby? And then one can also argue that the most recent settlement outposts, such as the resuscitated Homesh colony, damage the core business of the occupation and cannot be in its interest . . . and so on. We would here enter an infinite regress into the margins. Neither the US empire nor the Zionist project – to just stick with two top candidates – would have any endogenous driving forces, but rather slavishly follow the leash of those at the farthest remove from the centre. I don't know what would be left of anti-imperialism or anti-Zionism (or, in the next step, anti-capitalism) with such a procedure.

McNally reaches the finish line by alleging that a 'concrete analysis points toward Israel's increasing strategic superfluity to the American empire, and so suggests a heightened role for the lobby in ensuring continued sponsorship'. I find this to be an astounding statement. At the height of the transnationally organised genocide, Israel would possess *strategic superfluity* to the empire. The burden of proof the lobby theory would then have to carry is so large, so extreme as to beggar belief. This is not only a surrender of analytical judgment, but a farewell to Palestinian Marxism. On 18 July, Omar Murad, a member of the PFLP politburo, recapitulated the classical theory: the entity 'functions as a police officer and a colonial military base that ensures American and European imperialist hegemony and prevents the development, growth, and unity of Arab states' – the formula from 1840, now cause of 'genocide and ethnic cleansing'.[3] But McNally appears to imbue his version of the lobby theory with a progressive rationale of sorts. It is meant to help the left, namely the left in the US. It would be 'politically foolish' for it to nod along 'as Biden repeats that Israel is a trusty guarantor of American interests'. It is more prudent and effective to adopt American patriotism and claim that the US would be better off jettisoning Israel. This would presumably win more votes. And it might be in line with the social-democratic reformism of at least some *Jacobin* thinkers: but hardly with the foremost edge of the Palestine solidarity movement in the streets of the US. The first duty of a left in the heart of empire must be the ruthless critique of it. Posturing as the wisest interpreters of its true interests is not quite the best Leninist tradition.

But the stakes are higher still: McNally considers the 'stale notions of US empire as a frozen monolith' to be 'disempowering', particularly because such notions are 'often accompanied by grandiose rhetoric implying the Palestinians must await the toppling of Western civilization in its entirety for their deliverance from Zionism.'

We here reach a point where it becomes impossible not to think of the debate over the Holocaust. Was it the product of Western, bourgeois civilisation? Or was it an utter aberration from the West's generally good nature? The Frankfurt school, of course, came down very firmly on the former side. The capitalist West and the Zionist project found each other in the opposite view: the absolute exceptionality of the Holocaust. In his stellar essay 'Benjamin, the Holocaust and the Question of Palestine', in the important volume *The Holocaust and the Nakba: A New Grammar of Trauma and History*, Amnon Raz-Krakotzkin teases out the implications of that ideological unity. In the vein of Benjamin, he outlines a prophecy about the consequences of denying the Palestinians their rights *and* denying the tendency of Western, bourgeois civilisation to produce genocide. There will be repetition. There will be not identical, mechanical reiteration, but there will be a compulsion to destroy again. Denial 'can only end in destruction'.[4]

What brings another dimension to these debates – on top of, needless to say, the innumerable other differences between this genocide and the Holocaust or any other on the list – is climate. If the destruction of Palestine in at least some sense is the destruction of the earth, then, yes, the Palestinians are up against Western, bourgeois civilisation *in toto*. So is the rest of propertyless humanity. Whether that recognition is empowering or disempowering is a moot point. But if people marching on the streets of New York and elsewhere learn to hate the society that systematically turns the children of Gaza into ash and cinder, just as it does more and more regions of this planet, then I would not counsel them otherwise. They would be right to do so.

Notes

Preface: No Limits

1 Government Media Office, Gaza Strip, statistical update, 17 July 2024, *Middle East Observer*.
2 Frantz Fanon, *The Wretched of the Earth* (Broadway: Grove, 2004 [1963]), 17.
3 Kevin Liptak, 'Biden Says He Will Stop Sending Bombs and Artillery Shells to Israel if it Launches Major Invasion of Rafah', CNN, 9 May 2024.
4 Carlo Martuscelli, 'Biden Warns of "Red Line" for Israel over Gaza', *Politico*, 10 March 2024.
5 Stephen Collinson, 'Biden's Rafah Warning Sends Immediate Shockwaves through US and Global Politics', CNN, 9 May 2024.
6 The Mayor of Rafah, Dr Ahmed Al-Sufi, 24 July 2024, *Middle East Observer*.
7 'We're Protecting You: Full Text of Netanyahu's Address to Congress', *Times of Israel*, 25 July 2024; Jacob Magid, 'Netanyahu Checked All the Boxes on His US Trip – Except One', *Times of Israel*, 28 July 2024.
8 Oliver Milman and Nina Lakhani, 'Revealed: Wealthy Western Countries Lead in Global Oil and Gas Expansion', *Guardian*, 24 July 2024.

The Destruction of Palestine Is the Destruction of the Earth

1 The best primer on the genocide available in early April 2024, and still so in late July, is Francesca Albanese, 'Anatomy of a Genocide: Report of the Special Rapporteur on the Situation of Human Rights in the Palestinian Territories Occupied since 1967', United Nations, 25 March 2024.
2 Tom Stevenson, 'Rubble from Bone', *London Review of Books*, 8 February 2024.
3 See, for example, Nicholas R. Micinski and Kelsey Norman, 'Funding for Refugees Has Long Been Politicized – Punitive Action against UNRWA and Palestinians Fit this Pattern', *Conversation*, 1 February 2024. The new UK Labour government under Keir Starmer promised to resume funding of UNRWA in July 2024, while the US stood firm on the defunding.
4 As documented in, to pick two mainstream news outlets, Josh Holder, 'Gaza after Nine Weeks of War', *New York Times*, 12 December 2023; Niels de Hoog, Antonio Voce, Elena Morresi et al., 'How War Destroyed Gaza's Neighbourhoods – Investigation', *Guardian*, 30 January 2024.
5 Jared Malsin and Saeed Shah, 'The Ruined Landscape of Gaza after Nearly Three Months of Bombing', *Wall Street Journal*, 30 December 2023.
6 Albanese, 'Anatomy of a Genocide', 1, 11.
7 Plan Dalet quoted in Ilan Pappe, *The Ethnic Cleansing of Palestine*, (Oxford: Oneworld, 2007), 82; see further, 64, 77–8, 88, 147.
8 This and some other arguments in the text recycle material from Andreas Malm, 'The Walls of the Tank: On Palestinian Resistance', *Salvage*, 1 May 2017.

9 Liyana Badr, *A Balcony over the Fakihani* (New York: Interlink Books, 2002 [1993]), 73, 76, 81.

10 For example, Thomas E. Lovejoy and Carlos Nobre, 'Amazon Tipping Point: Last Chance for Action', *Science Advances* (2019) 5: 1–2; Chris A. Boulton, Timothy M. Lenton and Niklas Boers, 'Pronounced Loss of Amazon Rainforest Resilience since the Early 2000s', *Nature Climate Change* (2022) 12: 271–8; James S. Albert, Ana C. Carnaval, Suzette G. A. Flantua et al., 'Human Impacts Outpace Natural Processes in the Amazon', *Science* (2023) 379: 1–10; Meghie Rodrigues, 'The Amazon's Record-Setting Drought: How Bad Will It Be?', *Nature* (2023) 623: 675–6; and for further documentation and discussion, Wim Carton and Andreas Malm, *The Long Heat: Climate Politics When It's Too Late* (London: Verso, 2025).

11 Patrick Wintour and Luke Harding, '"Sea Is Constantly Dumping Bodies": Fears Libya Flood Death Toll May Hit 20,000', *Guardian*, 13 September 2023.

12 Jawhar Ali in Mohammed Abdusamee, Vivian Nereim and Isabella Kwai, 'More Than 5,000 Dead in Libya as Collapsed Dams Worsen Flood Disaster', *New York Times*, 12 September 2023.

13 Raed Qazmouz in Ayman Nobani, '"In Derna, Death Is Everywhere": Palestinian Mission to Libya', Al Jazeera, 21 September 2023.

14 Mariam Zachariah, Vassiliki Kotroni, Lagouvardos Kostas et al., 'Interplay of Climate Change-Exacerbated Rainfall, Exposure and Vulnerability Led to Widespread Impacts in the Mediterranean Region', World Weather Attribution, Imperial College London, 18 September 2023.

15 Nidal Al-Mughrabi, 'Palestinian Family That Fled Wars Suffers Death in Libya', *Reuters*, 14 September 2023.

16 Bethan McKernan and Quique Kierszenbaum, '"We're Focused on Maximum Damage": Ground Offensive into

Gaza Seems Imminent', *Guardian*, 10 October 2023; Revital Gottlieb, Likud MK, quoted on Yehuda Shaul, X account, 17 October 2023.

17 Victoria Bekiempis, 'Derek Chauvin Trial: Jury Begins Deliberations over Killing of George Floyd – as It Happened', *Guardian*, 20 April 2021.

18 Charles Napier, *The Navy: Its Past and Present State* (London: John & Daniel A. Darling, 1851), 48. Note that only a minimum of references – mostly the sources of direct quotations – is included in what follows.

19 F. S. Rodkey, 'Colonel Campbell's Report on Egypt in 1840, with Lord Palmerston's Comments', *Cambridge Historical Journal* (1929) 3: 112.

20 Hansard, House of Commons, vol. 49, 6 August 1839, 1391–2.

21 Quoted in C. K. Webster, *The Foreign Policy of Palmerston, 1830–41: Britain, the Liberal Movement and the Eastern Question* (London: Bell, 1951), 629.

22 Colonel Hodges quoted in William Holt Yates, *The Modern History and Condition of Egypt, Vol. 1* (London: Smith, Elder and Co., 1843), 428 (emphasis in original).

23 Broadlands Archive, University of Southampton: Lord Ponsonby quoted in 'Constantinople 22 March 1846: Secret Memorandum on the Syrian War of 1840–1841', by General Jochmus, MM/SY/1-3.

24 David K. Brown, *Before the Ironclad: Development of Ship Design, Propulsion and Armament in the Royal Navy, 1815–60* (London: Conway Maritime Press, 1990), 61.

25 Letter from Charles Napier to Colonel Hodges, 23 August 1840, in Elers Napier, *The Life and Correspondence of Admiral Sir Charles Napier, Vol. II* (London: Hurst and Blackett, 1862), 21 (emphasis in original).

26 Quoted in W. P. Hunter, *Narrative of the Late Expedition to Syria, Vol. I* (London: Henry Colburn, 1842), 69–70.

27 Quoted in Letitia W. Ufford, *The Pasha: How Mehmet Ali Defied the West, 1839–1841*(Jefferson: McFarland, 2007), 141.
28 Letter sent 25 September in Charles Napier, *The War in Syria, Vol. I* (London: John W. Parker, 1842), 83, 124.
29 Broadlands Archive: Lord Palmerston to Lord Ponsonby, 5 October 1840, GC/PO/755-769.
30 *The Mirror of Literature, Amusement and Instruction*, 'Burford's Panorama', 13 February 1841, 107 (emphasis in original).
31 Napier, *The War*, 206.
32 Robert Burford, *Description of a View of the Bombardment of St Jean D'Acre* (London: Geo. Nichols, 1841), 8, 3.
33 Captain Henderson quoted in Yates, *The Modern History*, 435.
34 Elliot Papers, National Maritime Museum: Lord Minto to Robert Stopford, 7 October 1840, ELL/216.
35 Report by colonel Charles F. Smith to Lord Palmerston in 'Correspondence Relative to the Affairs of the Levant', Parliamentary Papers, 1841, VIII, 56.
36 *Tait's Edinburgh Magazine for 1841*, 'Political register', 1841, VIII, 65.
37 Letter from Charles Napier to Eliza Napier, 13 November 1840, included in Napier, *The Life and Correspondence*, 113.
38 Napier, *The War*, 211.
39 Elliot Papers: Robert Stopford to Lord Minto, 5 November 1840, ELL/214. Stopford was the top British commander during the battle at Akka.
40 Account of Mr Hunt in W. P. Hunter, *Narrative of the Late Expedition to Syria, Vol. I* (London: Henry Colburn, 1842), 310.
41 Yaacov Kahanov, Eliezer Stern, Deborah Cvikel and Yoav Me-Bar, 'Between Shoal and Wall: The Naval

Bombardment of Akko, 1840', *Mariner's Mirror* (2014) 100: 160.

42 Letter from H. J. Codrington to E. Codrington, 4 November 1840, in *Selections from the Letters (Private and Professional) of Sir Henry Codrington* (London: Spottiswoode & Co, 1880), 162.

43 Broadlands Archive: Lord Palmerston to Lord Ponsonby, 14 November 1840, GC/PO/755-769.

44 Yates, *The Modern History*, 474.

45 Letter from General Jochmus to Lord Ponsonby, 17 January 1841, in *August von Jochmus' Gessamelte Schriften, Erster Band: The Syrian War and the Decline of the Ottoman Empire, 1840–1848* (Berlin: Albert Cohn, 1883), 84 (cf. 178).

46 *Tait's*, 'Political Register', 65.

47 'Iron War Steamers', *Manchester Guardian*, 14 April 1841.

48 'The Recent Victories', *Observer*, 28 November 1842.

49 John Bowring, *Report on Egypt and Candia: Addressed to the Right Hon. Lord Viscount Palmerston* (London: W. Clowes and Sons, 1840), 147.

50 A. A. Paton, *A History of the Egyptian Revolution, Vol. II* (London: Trübner & Co., 1863), 239.

51 For some further context and references, see Andreas Malm and the Zetkin Collective, *White Skin, Black Fuel: On the Danger of Fossil Fascism* (London: Verso, 2021), 343–63.

52 Forthcoming from Verso.

53 Broadlands Archive: Lord Palmerston to Lord Ponsonby, 25 November 1840, GC/PO/755-769.

54 Broadlands Archive: Lord Ashley (later Earl of Shaftesbury) to Lord Palmerston, 19 April 1836, GC/SH/2-22. The commercial potential of Palestine was also highlighted in another more extensive report to Lord Palmerston: John Bowring, *Report on the Commercial*

Statistics of Syria, addressed to the Right Hon. Lord Viscount Palmerston (London: William Clowes and Sons, 1840), for example 14–19, 30.

55 Quoted in Eitan Bar-Yosef, 'Christian Zionism and Victorian Culture', *Israel Studies* (2003) 8: 28.

56 Lady Palmerston on 3 December 1841, in Tresham Lever, *The Letters of Lady Palmerston* (London: John Murray, 1957), 243–4 (emphasis in original).

57 Broadlands Archive: Lord Palmerston to Lord Auckland, 22 January 1841, GC/AU/63/1.

58 Quoted in, for example, Regina Sharif, 'Christians for Zion, 1600–1919', *Journal of Palestine Studies* (1976) 5: 130; Herbert A. Yoskowitz, 'British Zionistic Writings Revisited', *European Judaism* (1979) 13: 45; Shlomo Sand, *The Invention of the Land of Israel: From Holy Land to Homeland* (London: Verso, 2012), 153.

59 First two letters quoted in Sharif, 'Christians for Zion', 130; Bar-Yosef, 'Christian Zionism', 29; third: Broadlands Archive: Lord Palmerston to Lord Ponsonby, 4 December 1840, GC/PO/755-769.

60 *The Times*, 17 August 1840.

61 'Syria', *Morning Herald*, 3 May 1841.

62 Quoted in Sharif, 'Christians for Zion', 132.

63 Colonel Churchill to Sir Moses Montefiore, 14 June 1841, in Lucien Wolf, *Notes on the Diplomatic History of the Jewish Question, with Texts of Treaty Stipulations and other Official Documents* (London: Spottiswoode, Ballantyne & Co., 1919), 119–21 (emphasis in original).

64 Quoted in *The Voice of Israel*, 'The Tranquilization of Syria and the East', 1 September 1845, 168 (both italics and capitalised letters in original).

65 Quoted in Albert M. Hyamson, 'British Projects for the Restoration of Jews to Palestine', *Publications of the American Jewish Historical Society* (1918) 26: 143.

66 Quoted in Sharif, 'Christians for Zion', 131.

67 'Christian Zionist Hall of Fame: Lord Shaftesbury', *Israel Answers*, 2024.

68 Diana Muir, '"A Land without a People for a People without a Land"', *Middle East Quarterly*, Spring 2008.

69 Alexander Keith, *The Land of Israel, according to the Covenant with Abraham, with Isaac, and with Jacob* (Edinburgh: William Whyte & Co., 1843), 34, 382, 366.

70 Ibid., 382 (emphases in original).

71 Quoted in Bar-Yosef, 'Christian Zionism', 29.

72 *Morning Post*, 'The Jews', 30 January 1841.

73 Anon., *'The Kings of the East': An Exposition of the Prophecies Determining, from Scripture and from History, the Power for Whom the Mystical Euphrates Is Being 'Dried Up'; with an Explanation of Certain Other Prophecies Concerning the Restoration of Israel* (London: R. B. Seeley and W. Burnside, 1842), 277; on steam as pillar of power, see 48–50.

74 Ibid., 209, 211 (report from the *Times*).

75 Ibid., 212.

76 Ibid., 204–6.

77 Ibid., 212 (italics in original).

78 *The Western Messenger*, 'Restoration of the Jews to Palestine', October 1840, 264, 266.

79 On this status of Noah, see Louis Ruchames, 'Mordecai Manuel Noah and Early American Zionism', *American Jewish Historical Quarterly* (1975) 64: 195–223. Coincidentally or not, Noah was also 'among the most prominent opponents of the abolition of slavery, using his position as editor of the *New York Evening Star* to characterize African-Americans as mentally inferior to whites, to support the so-called gag rule preventing the Senate from discussing slavery, and even to argue for a move "to make publication of antislavery literature a punishable offence."' Joseph Phelan, '"How Came They Here?": Longfellow's

"The Jewish Cemetary at Newport", Slavery, and Proto-Zionism', *EHL* (2020) 87: 141.

80 M. M. Noah, *Discourse on the Restoration of the Jews* (New York: Harper & Brothers, 1845), 10, 35–6.

81 Ibid., 47–8.

82 Ibid., 39.

83 Ibid., 35.

84 Ibid., 38.

85 Jonathan Parry, *Promised Lands: The British and the Ottoman Middle East* (Princeton: Princeton University Press, 2022), 376.

86 Ibid., 143.

87 Ibid., 15.

88 Yoskowitz, 'British Zionistic', 45.

89 Albanese, 'Report of the Special', 2.

90 Quoted in ibid., 14.

91 Parry, *Promised Lands*, 13.

92 'The Jews', *National Repository*, March 1877, 274.

93 Ilan Pappe, *The Ethnic Cleansing of Palestine*, 93.

94 Eliahu Epstein quoted in Irene L. Gendzier, *Dying to Forget: Oil, Power, Palestine, and the Foundations of US Policy in the Middle East* (New York: Columbia University Press, 2015), 105.

95 Paul Thomas Chamberlin, *The Global Offensive: The United States, the Palestine Liberation Organization, and the Making of the Post–Cold War Order* (Oxford: Oxford University Press, 2015), 138.

96 Quoted in ibid., 125.

97 Oliver Milman, 'Surge of New US-Led Oil and Gas Activity Threatens to Wreck Paris Climate Goals', *Guardian*, 28 March 2024.

98 'Israel and Germany Sign Deal for Ships to Guard Gas Rigs', *Times of Israel*, 11 May 2015.

99 Sarah El Safty and Ari Rabinovitch, 'EU, Israel and Egypt Sign Deal to Boost East Med Gas Exports to Europe',

Reuters, 15 June 2022; *Times of Israel*, 'Israel Exports Crude Oil for First Time, with Shipment Heading for Europe', 16 February 2023; Steven Scheer, 'Energean Starts Gas Production at Israel's Karish Site', *Reuters*, 26 October 2022.

100 Anna Cooban and Matt Egan, 'Israel Just Shut a Gas Field Near Gaza. Here's Why That Matters', CNN, 10 October 2023.

101 Stanley Reed, 'Chevron Shuts Down Natural Gas Platform Near Gaza Strip', *New York Times*, 9 October 2023.

102 Ron Bousso and Sabrina Valle, 'Chevron Resumes Natural Gas Supply from Israel's Tamar Offshore Field', *Reuters*, 13 November 2023.

103 Sharon Wrobel, 'Chevron Partners Greenlight $24m Investment to Boost Gas Production at Offshore Site', *Times of Israel*, 18 February 2024.

104 Ari Rabinovitch and Steven Scheer, 'Israel Awards Gas Exploration Licenses to Eni, BP and Four Others', *Reuters*, 30 October 2023.

105 Kjell Kühne, Nils Bartsch, Ryan Driskell Tate et al., '"Carbon Bombs": Mapping Key Fossil Fuel Projects', *Energy Policy* (2022) 166: 1–10; Scott Zimmerman and Hanna Fralikhina, 'Hooked on Hydrocarbons: The UK's Risky Addiction to North Sea Oil and Gas', *Global Energy Monitor*, October 2022; 'Ithaca Energy Acquires Shell's Stake in UK Cambo Field', *Reuters*, 12 September 2023; Sarah Young, 'Britain Gives Go-Ahead for Biggest New North Sea Oilfield in Years', *Reuters*, 27 September 2023; 'Total Energies Hits Hydrocarbons at North Sea Appraisal Well', *Offshore Engineer*, 4 January 2023.

106 'Ithaca Energy Set for London's Biggest IPO in 2022', *Reuters*, 2 November 2022.

107 Benjamin Neimark, Patrick Bigger, Frederick Out-Larbi and Reuben Larbi, 'A Multitemporal Snapshot of Green-

house Gas Emissions from the Israel–Gaza Conflict', Queen Mary University of London, 2024; cf. Nina Lakhani, 'Emissions from Israel's War in Gaza Have "Immense" Effect on Climate Catastrophe', *Guardian*, 9 January 2024.

108 Neta C. Crawford, *The Pentagon, Climate Change, and War: Charting the Rise and Fall of US Military Emissions* (Cambridge, MA: MIT Press, 2022), 7–8.

109 Ghada Karmi, *Married to Another Man: Israel's Dilemma in Palestine* (London: Pluto, 2007), 84.

110 Ibid., 91.

111 Ibid., 103.

112 Ibid., 97–8.

113 Sayyed Hassan Nasrallah, Al Manar TV, 3 September 2012, translated by Memri.

114 PFLP, *Strategy for the Liberation of Palestine* (Utrecht: Foreign Language Press, 2017), 34, 101, 102.

115 'Political Document of Palestinian Islamic Jihad', in Erik Skare (ed.), *Palestinian Islamic Jihad: Islamist Writings on Resistance and Religion* (London: I. B. Tauris, 2021 [2018]), 31–2.

116 Fatih al-Shiqaqi, 'The Palestinian Cause Is the Central Question of the Islamic Movement . . . Why?', in ibid. [1980], 77.

117 'Senate Session, 5 June 1986, Joe Biden: Were There Not an Israel the USA Would Have to Invent an Israel to Protect Her Interest in the Region', C-Span, 11 May 2021.

118 Yitzhak Benhorin, 'Biden in 2007 Interview: I Am a Zionist', Ynet, 23 July 2008; the White House, 'Remarks by Vice President Biden: The Enduring Partnership between the United States and Israel', 11 March 2010; the White House, 'Remarks by Vice President Joe Biden the 67th Annual Israeli Independence Day Celebration', 23 April 2015; M. Muhannad Ayyash, 'Biden Says That the

U.S. Would Have to Invent an Israel if It Didn't Exist. Why?', *Conversation*, 25 July 2023.

119 Pappe, *The Ethnic Cleansing of Palestine*, 261.

120 Quoted in D. A. Jaber, 'Settler Colonialism and Ecocide: Case Study of Al-Khader, Palestine', *Settler Colonial Studies* (2019) 9: 135.

121 For one example of particularly superb work, see Matan Kaminer, 'The Agricultural Settlement of the Arabah and the Political Ecology of Zionism', *International Journal of Middle Eastern Studies* (2022) 54: 40–56.

122 For the longer perspective, see Shourideh C. Molavi, *Environmental Warfare in Gaza: Colonial Violence and New Landscapes of Resistance* (London: Pluto, 2024); for the ecocidal war, UNEP, 'Environmental Impact of the Conflict in Gaza: Preliminary Assessment of Environmental Impacts', 2024; and cf. Kaamil Ahmed, Damien Gayle and Aseel Mousa, '"Ecocide in Gaza": Does Scale of Environmental Destruction Amount to a War Crime?', *Guardian*, 29 March 2024.

123 Seth J. Frantzman, 'How Did Israel Fail to Stop Hamas' October 7 Attack?', *Jerusalem Post*, 13 October 2023.

124 Michele Groppi and Vasco da Cruz Amador, 'Technology and Its Pivotal Role in Hamas' Successful Attacks on Israel', Global Network on Extremism and Technology, 20 October 2023.

125 PFLP, *Strategy*, 95.

126 And, it should be added, the Palestinian fighters had no firepower of the kind that could cause the massive destruction seen in some of the kibbutzes and along some roads.

127 Yonah Jeremy Bob, 'Gallant Warns Hezbollah: Israel Can Do in Beirut What It Is Doing in Gaza', *Jerusalem Post*, 17 December 2023; *Times of Israel*, 'Gallant Warns: If Hezbollah Isn't Deterred, Israel Can "Copy-Paste" Gaza War to Beirut', 8 January 2024.

128 Quoted in Yuval Abraham, '"A Mass Assassination Factory": Inside Israel's Calculated Bombing of Gaza', +972 *Magazine*, 30 November 2023.

129 Yuval Abraham, '"Lavender": The AI Machine Directing Israel's Bombing Spree in Gaza', +972 *Magazine*, 3 April 2024.

Response to Some Objections Regarding the Palestinian Resistance

1 Matan Kaminer, 'After the Flood: A Response to Andreas Malm', Verso Blog, 10 May 2024.

2 On the spirit of comradely critique, see Lukas Slothuus, 'Comradely Critique', *Political Studies* (2023) 71: 714–32.

3 '"For the Sake of the Front", a PFLP song released in 2022, debuting at a PFLP festival celebrating the 55th anniversary of its launch', Songs from the Resistance, X, 19 November 2023.

4 See 'Short thread of compiled media relating to the recently formed women's units of the National Resistance Brigades, the armed wing of the DFLP', Songs from the Resistance, X, 1 November 2023.

5 Andreas Malm and Shora Esmailian, *Iran on the Brink: Rising Workers and Threats of War* (London: Pluto, 2007); Andreas Malm, 'The Dialectics of Disaster: Considerations on Hazards and Vulnerability in the Age of Climate Breakdown, with a Brief Case Study of Khuzestan', *Jamba: Journal of Disaster Risk Studies* (2023) 15: 1–9.

6 Adi Callai, 'The Gaza Ghetto Uprising', *Brooklyn Rail*, May 2024.

7 The literature is already endless, but see particularly Al Jazeera Special Investigations, 'October 7', 20 March 2024; as well as the work of *Mondoweiss* and *Electronic Intifada*.

8 Nelson Mandela, *Long Walk to Freedom: The Autobiography of Nelson Mandela* (London: Abacus, 1995) 617–18.

9 Robert Windrem, 'US Government Considered Nelson Mandela a Terrorist Until 2008', NBC News, 7 December 2013.

10 'Hamba Kahle Umkhonto We Sizwe', YouTube, Blackpilled Mirpuri, 9 November 2020.

11 'Gaza: Nelson Mandela Receives Welcome from Assembly', YouTube, AP Archive, 20 October 1999.

12 See Hamas Media Office, 'Our Narrative ... Operation Al-Aqsa Flood', 2024, available from *Palestine Chronicle*.

13 Ali Abunimah, 'It's Time to Change Liberal Discourse about Hamas', *Electronic Intifada*, 10 June 2021.

14 Sara Roy, *Hamas and Civil Society in Gaza: Engaging the Islamist Social Sector* (Princeton: Princeton University Press, 2013); Tareq Baconi, *Hamas Contained: A History of the Palestinian Resistance* (Stanford: Stanford University Press, 2018); Somdeep Sen, *Decolonizing Palestine: Hamas between the Anticolonial and the Postcolonial* (Ithaca: Cornell University Press, 2020).

15 Erik Skare, *A History of Palestinian Islamic Jihad: Faith, Awareness, and Revolution in the Middle East* (Cambridge: Cambridge University Press, 2021).

16 Rachel Fink, 'Explained: The Israeli Extremists Who Want to Rebuild the Temple, and the Government Ministers Who Back Them', *Haaretz*, 24 April 2024; Ronen Bergman, 'The Unpunished: How Extremists Took Over Israel', *New York Times*, 16 May 2024.

17 By far the best study of the political culture of the martyrdom operations is Nasser Abufarha, *The Making of a Human Bomb: An Ethnography of Palestinian Resistance* (Durham: Duke University Press, 2009).

18 There will be 'swarms of suicide bombers' waiting for the IDF, predicted the senior international correspondent of CNN: Sam Kiley, 'Israel's Allies Fear It Could be Walking into a Trap in Gaza as Hamas and Its Backers Seek a Wider Conflict', CNN, 26 October 2023.
19 Yaniv Kubovich, 'Israeli Army Commanders Gave Order to Blow Up Rafah Reservoir. IDF Suspects Breach of Int'l Law', *Haaretz*, 29 July 2024.
20 See, for examle, Bethan McKernan, 'Nizar Banat's Death Highlights Brutality of Palestinian Authority', *Guardian*, 31 August 2021.
21 See, for example, Ylenia Gostoli, 'Basil al-Araj, Palestinian Activist, Buried in West Bank', Al Jazeera, 18 March 2017.
22 The joint operations room was highlighted in Abdeladi Ragad, Richard Irvine-Brown, Benedict German and Sean Seddon, 'How Hamas Built a Force to Attack Israel on 7 October', BBC, 27 November 2023.
23 Etan Nechin, '"Our Future Is Here": Christians in Gaza Are Paying an Unholy Price for Israel–Hamas War', *Haaretz*, 21 April 2024.
24 See, for example, Khaled Abu Toameh and Tovah Lazaroff, 'Hamas Demanding Release of Barghouti and Sa'adat in Prisoner Swap', *Jerusalem Post*, 7 May 2020.
25 See the website and particularly the podcast of *Electronic Intifada*, which includes the commentary by Jon Elmer, much of whose work can also be accessed through his X account.
26 Bashir Abu-Manneh, 'The Palestinian Resistance Isn't a Monolith', *Jacobin*, 28 April 2024.
27 On the Zionist hatred of the people of Gaza and early desires to empty the Strip, see the excellent Jean-Pierre Filiu, *Gaza: A History* (London: Hurst, 2014).

Response to Some Objections Regarding the Israel Lobby

1 Ed McNally, 'The Israel Lobby Matters', *Jacobin*, 8 May 2024.

2 On what West Germany did for Israel between 1948 and 1967 (and also, of course, beyond), see the excellent Daniel Marwecki, *Germany and Israel: Whitewashing and Statebuilding* (London: Hurst, 2020).

3 Omar Murad to Sawt al-Shaab Radio, Lebanon, 18 July 2024, *Resistance News Network*.

4 Amnon Raz-Krakotzkin, 'Benjamin, the Holocaust, and the Question of Palestine', in Amos Goldberg and Bashir Bashir (eds), *The Holocaust and the Nakba: A New Grammar of Trauma and History* (New York: Columbia University Press, 2019), 90.